MAMMALS

Laura Perdew

Abdo Reference

An Imprint of Abdo Publishing | abdobooks.com

CONTENTS

Insectivores

Bats

Primates

Hoofed Mammals

Elephants

Marsupials

Monotremes

WHAT IS A MAMMAL?

There are more than 5,000 species of mammals on Earth. They are a diverse group, yet they have several things in common:

- All mammals are vertebrates—they have a backbone.
- They are also warm-blooded, which means they produce their own body heat and do not have to rely on the sun or outside temperature to stay warm.
- All mammals have hair. For some mammals, the hair helps to keep them warm. Fur colors or patterns help to camouflage some mammals in their environments, keeping them safe from predators.
- Almost all mammals give birth to live young. The only exceptions are platypuses and echidnas. These two mammals lay eggs.

WHERE ARE MAMMALS FOUND?

Mammals are found on every continent and in every ocean. They are considered some of the planet's most adaptable creatures. They live in a wide range of habitats, including the cold and icy Arctic regions; the arid, hot deserts of Africa; rain forests; and mountaintops. In fact, mammals are found in every land habitat on Earth. Some of these include:

- Above or below ground, and some mammals split their time between both.
- Some mammals live their whole lives in trees.
- Other mammals spend their whole lives in the ocean or near the ocean.

WHAT CAN MAMMALS DO, AND WHAT ARE THEY LIKE?

Mammals can run, jump, fly, hop, pounce, climb, swing, burrow, swim, walk, dive, and scurry. They vary in size—from the tiny American pygmy shrews and bumblebee bats, to the elephants and enormous blue whales. Many male mammals are larger than their female counterparts. Mammals have diverse diets too. Many are carnivores, while others are herbivores, omnivores, or insectivores.

The behavior of mammals differs also. Some mammals live alone throughout most of their lives. They may only spend time with others of their species during mating season or when they are young or raising young. Other mammals live in highly organized social groups. In addition, some species spend time hunting and moving during the day. These mammals are known as diurnal. Others are nocturnal, or only active at night. And some are crepuscular, which means they are only active at dusk and dawn.

HOW TO USE THIS BOOK

Tab shows the mammal category.

DOG FAMILY (CANINES)

GRAY WOLF *(CANIS LUPUS)*

The coats of gray wolves are often gray, but they can also be black, reddish, brown, or white. They are the largest of all canines and are known for their distinctive howls, which are ...ey communicate with one another. They are highly ...ving in packs of two to 15 individuals, usually family ...rs. Each pack has a strict social order, works together ... the pups, and coordinates hunts.

The mammal's common name appears here.

...W TO SPOT

Size: 28 to 31 inches (71 to 79 cm) at the shoulder; tail length is 13 to 20 inches (33 to 51 cm); 40 to 175 pounds (18 to 80 kg)

Range: Alaska, western and northeastern Canada, northwestern United States, and certain areas of Europe and Asia

...abitat: Forests and tundra

...arge mammals, including ...eer, moose, and caribou, as ...s hares and beavers

Fun Facts give interesting information about mammals.

FUN FACT

Gray wolves that live near the coast in Canada will swim between islands and will eat seafood, including crabs and clams.

DOMESTIC DOGS

... were the first animal to be domesticated by humans. ...ns used dogs to help with hunting and herding, as ...s, and for companionship. Today, there are hundreds ...mestic dog breeds, but they all share a common ...stor: the gray wolf.

32

Sidebars provide additional information about the topic.

The mammal's scientific name appears here.

RCTIC FOX *(VULPES LAGOPUS)*

This paragraph gives information about the mammal.

ctic foxes are small canines with white fur in the wint
d gray-brown fur in the summer. This allows them to
o the habitat as the seasons change. Other adapta
at help them survive the Arctic cold are compact bo
nall ears, and thick fur. They also create dens with a
ystem of tunnels that are used generation after gene
nd can have dozens of entrances.

HOW TO SPOT

Size: 12 inches (30 cm) at the shoulder; tail length is up to 13.75 inches (34.9 cm); 6.5 to 17 pounds (2.9 to 7.7 kg)

Range: Arctic regions of Europe, Asia, North America, Greenland, and Iceland

Habitat: Arctic tundra, generally near coastlines

Diet: Small rodents, Arctic hares, and birds

How to Spot features give information about the mammal's size, range, habitat, and diet.

33

Images show the mammal.

NARWHAL *(MONODON MONOCEROS)*

Narwhals have a single tusk, which is actually a tooth. It grows out of the upper left jaw of male narwhals in a counterclockwise spiral. The tusks can grow to more than 10 feet (3 m) long. Only very rarely will adult females grow a tusk. Sometimes, a narwhal grows two.

MARINE MAMMALS

Marine mammals live all or most of their lives in or near the ocean. They depend on the ocean for their survival. In addition, these mammals have a thick layer of blubber to help them keep warm. Marine mammals are found all over the world in a variety of habitats.

HOW TO SPOT

Size: 13 to 16 feet (4 to 5 m) long; 2,200 to 4,000 pounds (1,000 to 1,800 kg)

Range: Arctic Ocean and north of the Arctic Circle

Habitat: Depends on the season. They can be found in the deep ocean, ice-free coastal areas, and densely packed ice.

Diet: Greenland halibut and other fish, squid, and crustaceans

BELUGA *(DELPHINAPTERUS LEUCAS)*

Beluga whales have all-white skin and a melon-shaped head. Belugas are highly social creatures and travel in large pods. They are very vocal too, using a variety of clicks, whistles, chirps, and squawks for communicating, mating, and hunting.

HOW TO SPOT

Size: 13 to 16 feet (4 to 5 m) long; up to 3,500 pounds (1,600 kg)

Range: Arctic and subarctic waters

Habitat: Shallow, coastal waters of the Arctic seas and in deep waters

Diet: A variety of fish, crustaceans, and mollusks

FUN FACT

Both narwhals and belugas use echolocation to hunt in the ocean.

BOTTLENOSE DOLPHIN

(TURSIOPS TRUNCATUS)

Bottlenose dolphins are highly intelligent and use a wide variety of clicks, grunts, whistles, squeaks, and whines to communicate. They travel in pods of two to more than 100 dolphins, often forming pair bonds that last many years. Bottlenose dolphins have thick, short snouts and gray skin.

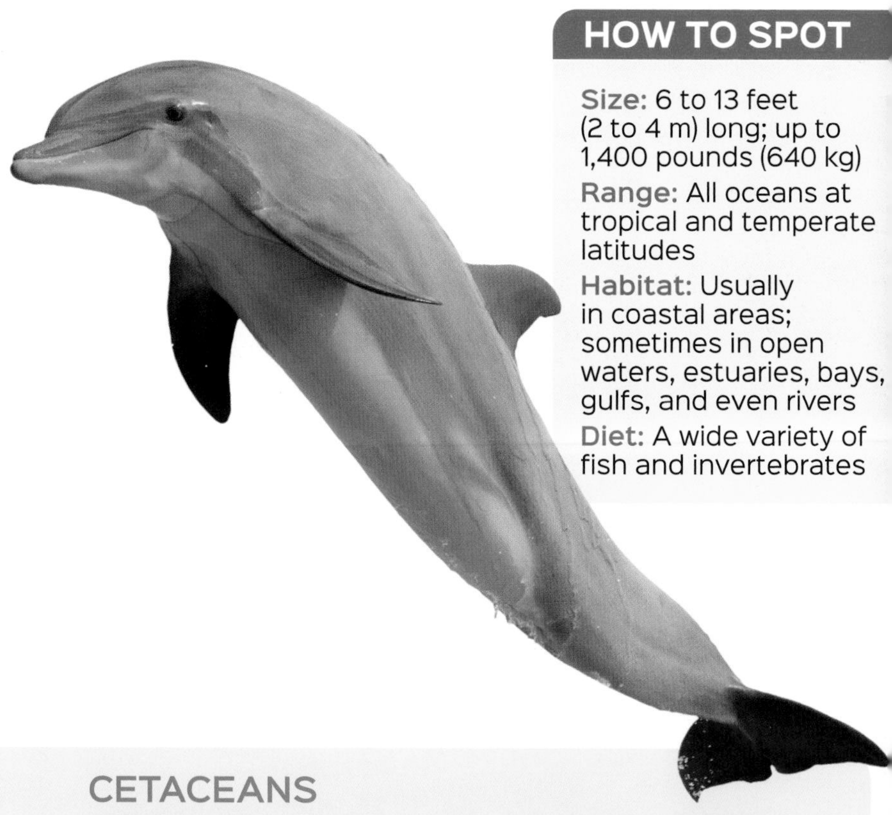

HOW TO SPOT

Size: 6 to 13 feet (2 to 4 m) long; up to 1,400 pounds (640 kg)

Range: All oceans at tropical and temperate latitudes

Habitat: Usually in coastal areas; sometimes in open waters, estuaries, bays, gulfs, and even rivers

Diet: A wide variety of fish and invertebrates

CETACEANS

Whales, dolphins, and porpoises are a type of marine mammal called cetaceans. Cetaceans are divided into two groups: toothed and baleen whales. As the name suggests, toothed whales (belugas, narwhals, dolphins, and porpoises) have teeth and one blowhole. Baleen whales have two blowholes and no teeth. Instead, they have hundreds of baleen plates that trap prey when the whales take water into their mouths.

HARBOR PORPOISE

(PHOCOENA PHOCOENA)

Harbor porpoises are shy, solitary animals. They usually prefer to travel alone or in small groups of two to five others. They tend to avoid boats and humans. Harbor porpoises are mobile. They may travel many miles per day over a home range that can extend thousands of square miles. Harbor porpoises have rounded faces and short beaks. Their throats and stomachs are white, and their backs are dark gray.

HOW TO SPOT

Size: 5 to 5.5 feet (1.5 to 1.7 m) long; 135 to 170 pounds (61 to 77 kg); females are larger than males

Range: Coastal areas of the North Atlantic, Arctic, and North Pacific Oceans, as well as the Black and Baltic Seas

Habitat: Shallow, coastal waters, including bays and estuaries; fresh water at the mouth of rivers

Diet: Fish, octopuses, and squid

FUN FACT
Underwater noise made by people can disrupt porpoises' communication.

Harbor porpoises have a single dorsal fin.

ORCA WHALE *(ORCINUS ORCA)*

Orca whales are a type of toothed whale. They have the nickname "killer whales" because they are fierce predators at the top of the marine food chain. They are very intelligent, often hunting in packs. In addition to intelligence, orcas exhibit complex social structures, and young are taught hunting and parenting skills.

HOW TO SPOT

Size: 26 to 30 feet (8 to 9 m) long; 8,400 to 12,000 pounds (3,800 to 5,400 kg)

Range: All the world's oceans

Habitat: Open oceans and coastal waters

Diet: A wide variety, including seals, sea lions, smaller whales and dolphins, fish, sharks, squid, octopuses, sea turtles, and otters

Orcas are the largest members of the dolphin family.

BLUE WHALE

(BALAENOPTERA MUSCULUS)

Blue whales, a type of baleen whale, are the largest animals on Earth. Despite their sizes, they feed almost exclusively on krill, a tiny marine organism. These whales are even enormous at birth, averaging 23 feet (7 m) long and weighing 5,000 to 6,000 pounds (2,270 to 2,720 kg). To sustain their calves, mother blue whales produce more than 50 gallons (190 L) of milk a day for six months.

FUN FACT

The heart of a blue whale is the size of a small car.

HOW TO SPOT

Size: 70 to 90 feet (21 to 27 m) long; 200,000 to 300,000 pounds (91,000 to 136,000 kg); males are smaller than females

Range: Every ocean except the Arctic

Habitat: Open oceans

Diet: Krill

CALIFORNIA SEA LION

(ZALOPHUS CALIFORNIANUS)

California sea lions are highly intelligent, playful, and vocal. They "bark" to communicate with one another. Female and young California sea lions range from tan to blond. Male adults have black or dark-brown skin.

HOW TO SPOT

Size: 6 to 7.5 feet (1.8 to 2.3 m) long; 240 to 700 pounds (110 to 320 kg)

Range: Pacific coast from south Alaska to Mexico

Habitat: Coastal waters of the Pacific and on beaches, docks, jetties, and buoys

Diet: Anchovies, squid, sardines, mackerel, and other fish

SOUTHERN ELEPHANT SEAL

(MIROUNGA LEONINE)

Southern elephant seals are the largest type of seal. They get their name from the trunk-like nose males develop. This elongated snout with an inflatable nose is used by males to vocalize threats to other males over territory or females. Their calls and vicious fights are used to establish social ranking.

Male

FUN FACT

While elephant seals look clumsy on land, they are skilled, agile swimmers. They can dive to depths of 5,000 feet (1,500 m) and stay underwater for as long as two hours.

HOW TO SPOT

Size: 10 to 16.5 feet (3 to 5 m) long; 2,000 to 11,000 pounds (900 to 5,000 kg); males are much heavier than females

Range: Antarctic and subantarctic waters to the southern coast of Argentina

Habitat: They breed on land but spend most of their time in open Antarctic waters.

Diet: Large fish and squid

Juvenile

Female

HAWAIIAN MONK SEAL

(NEOMONACHUS SCHAUINSLANDI)

Unlike other seals, Hawaiian monk seals live mostly alone. These seals have a thick skin fold around their necks. Adults have backs that are dark gray and bellies that are lighter in color. Their overall population began to decline in the early 1800s. This was in part because of commercial hunting. The Hawaiian monk seal has been considered an endangered species for many years.

HOW TO SPOT

Size: 6 to 7 feet (1.8 to 2.1 m) long; 400 to 600 pounds (180 to 270 kg); females are slightly bigger than males

Range: Northwestern Hawaiian Islands

Habitat: Warm, subtropical waters near certain islands, and offshore near reefs and submerged banks

Diet: Reef fish, squid, octopuses, and crustaceans

FUN FACT

Hawaiian monk seals can sleep for days at a time on beaches.

WALRUS *(ODOBENUS ROSMARUS)*

Walrus prefer to be near ice. They have tusks, which are actually long teeth. Sometimes they use their tusks to haul themselves out of the water. Males also use their tusks in battles over territory or to attract females. The winner of these fierce battles determines who is able to mate with a group of females.

HOW TO SPOT

Size: 10 to 12 feet (3 to 3.7 m) long; 2,600 to 4,200 pounds (1,200 to 1,900 kg)

Range: Across the Arctic region

Habitat: Shallow waters and coastlines near ice

Diet: Mussels, clams, and sometimes fish

WEST INDIAN MANATEE

(TRICHECHUS MANATUS)

West Indian manatees mostly eat plants. They have highly flexible lips with thick bristles that are able to wrap around and grasp plants on the seafloor. Then they use their lips to move the vegetation into their mouths. These mammals have wrinkled skin and can range in color from light brown to black. Their tails are shaped like paddles.

FUN FACT

Because of their sizes and diets, manatees must graze for six to eight hours a day. Their diets are also the reason for their nickname, "sea cows."

HOW TO SPOT

Size: 12 to 13.5 feet (3.7 to 4.1 m) long; up to 3,300 pounds (1,500 kg); females are larger than males

Range: Warm water of the western Atlantic Ocean

Habitat: Seagrass beds that are usually close to shore, and sometimes in or near freshwater sources

Diet: Seagrasses and other vegetation

DUGONG *(DUGONG DUGON)*

Dugongs are close cousins to manatees. They are herbivores, feeding on vegetation on the seafloor. The end of their large muzzles are broad, flattened, and turned downward. Dugongs are adapted to bottom-feeding. Like their cousins, they have flexible lips to move food to their mouths. They must eat large quantities of food every day due to their sizes.

HOW TO SPOT

Size: 11 to 13.5 feet (3.3 to 4.1 m) long; 2,200 pounds (1,000 kg)

Range: Tropical and subtropical waters of the Indian and western Pacific Oceans

Habitat: Shallow waters along coastlines; seagrass beds

Diet: Seagrasses

AMERICAN BLACK BEAR

(URSUS AMERICANUS)

Black bears are solitary, intelligent creatures that are common across the North American continent. These bears vary greatly in size, color, and diet depending on where they live. They are very adaptable, living in a variety of forest types and even in suburban areas.

HOW TO SPOT

Size: Black bear sizes vary depending on location, season, and the availability of quality food. They average 4 to 6 feet (1.2 to 1.8 m) long and weigh 88 to 660 pounds (40 to 300 kg).

Range: Across North America, from Canada to northern Mexico

Habitat: Forested areas

Diet: A wide variety of food depending on what is available, including nuts, berries, insects, small deer and moose, and salmon

FUN FACT

Even though they are called black bears, their fur can be black, blond, cinnamon, grayish, or white.

BROWN BEAR *(URSUS ARCTOS)*

In North America, brown bears are also known as grizzly bears. Their fur can be light to dark brown, and they have a hump on their backs between their shoulders. Brown bears are stocky, with powerful arms and sharp front claws that are 3 to 4 inches (7.6 to 10 cm) long.

HOW TO SPOT

Size: These mammals vary in size depending on location and food availability. They average 5 to 7 feet (1.5 to 2 m) long and weigh 200 to 1,000 pounds (90 to 450 kg).

Range: Europe, northern Asia, parts of the Himalayas, Japan, and the northwestern parts of North America

Habitat: Open areas, including tundra and coastlines. They're also found in alpine meadows and forests.

Diet: Many different things, including nuts, fruit, birds, fish, and even large hoofed mammals

GIANT PANDA

(AILUROPODA MELANOLEUCA)

Also called panda bears, these bears are easily recognized because of their contrasting black-and-white colorings. Giant pandas have black fur on their shoulders, legs, ears, and around their eyes. White fur covers the rest of them. They have stubs for tails, and they have the typical body shape of bears.

FUN FACT

For a long time, scientists thought pandas were raccoons because they share many characteristics. However, DNA testing has confirmed that pandas are indeed bears.

HOW TO SPOT

Size: 4 to 6 feet (1.2 to 1.8 m) long; 155 to 275 pounds (70 to 125 kg)

Range: Southwest China

Habitat: High mountain forests at altitudes between 4,000 to 13,400 feet (1,200 to 4,100 m) above sea level

Diet: Branches, stems, and leaves of bamboo

POLAR BEAR *(URSUS MARITIMUS)*

Polar bears are one of the largest types of bear. In addition, they are considered marine mammals because they are completely dependent on the ocean and sea ice for survival. Polar bears use ice edges and fractures in the ice as their hunting grounds. Due to global warming, polar bears' survival is threatened because the Arctic ice they depend on is thinning and melting earlier each year.

HOW TO SPOT

Size: 6.5 to 8.5 feet (2 to 2.6 m) long; 660 to 1,800 pounds (300 to 800 kg)

Range: Throughout the Arctic region in coastal areas

Habitat: Ice packs

Diet: Primarily seals

A polar bear's fur isn't really white. It's transparent, and it can change color depending on the angle of light and the season.

PUMA *(PUMA CONCOLOR)*

Pumas are also commonly known as mountain lions or cougars. These cats have solid-colored coats that range from yellowish to gray brown, with light-colored throats and chests. Pumas are extremely stealthy, powerful hunters. A puma can take down prey two to three times its size.

HOW TO SPOT

Size: 2.8 to 5 feet (0.8 to 1.5 m) long; tail length is 2 to 3 feet (0.6 to 0.9 m); 75 to 160 pounds (34 to 73 kg)

Range: Western Canada and the United States, Mexico, Central and South America

Habitat: Mountain and tropical forests, prairies, deserts, and swamps

Diet: Their diets depend on location. They can eat deer, pigs, capybaras, raccoons, armadillos, hares, squirrels, elk, and moose.

OCELOT *(LEOPARDUS PARDALIS)*

Ocelots are small cats with amazing patterns of dark spots on golden fur, and two dark stripes on each side of their faces. Like some other cats, if ocelots don't finish a meal in one sitting, they will hide it and return to it the next night.

HOW TO SPOT

Size: 2.4 to 3.25 feet (0.7 to 1 m) long; tail length is 0.8 to 1.3 feet (0.2 to 0.4 m); 14 to 35 pounds (6 to 16 kg)

Range: Mostly in Central and South America

Habitat: Areas with thick vegetation, including rain forests, mangrove forests, and thorn scrub areas

Diet: Mostly small rodents, but also birds, snakes, lizards, rabbits, deer, and fish

FUN FACT

There are close to 40 different species of wildcats in the world.

CANADIAN LYNX *(LYNX CANADENSIS)*

Canadian lynx have long, black tufts of hair on their ears. They also have long legs and enormous paws. These paws act like snowshoes, allowing lynx to walk on deep snow. Snowshoe hares are their main prey, and lynx populations rise and fall with snowshoe hare populations. Lynx hunt at night because that is when the hares are most active.

HOW TO SPOT

Size: 2.4 to 3.5 feet (0.7 to 1.1 m) long; tail length is 2 to 5 inches (5 to 13 cm); 11 to 38 pounds (5 to 17 kg)

Range: Alaska, across Canada, and the northern Rocky Mountains of the continental United States

Habitat: Thick boreal forests with high snowshoe hare populations

Diet: Primarily snowshoe hares

BOBCAT *(LYNX RUFUS)*

Bobcats are a species of lynx and are often misidentified. The black tufts of hair on a bobcat's ears are shorter than a lynx's, and a bobcat's tail has black stripes. In addition, they have smaller feet and are not adapted to living in snowy areas. They are, however, adapted to living in a wide range of other habitats and prefer places with dense cover or uneven terrain.

HOW TO SPOT

Size: 2 to 3.9 feet (0.6 to 1.2 m) long; tail length is 3 to 10 inches (7.6 to 25.4 cm); 13 to 44 pounds (6 to 20 kg)

Range: Across North America, from southern Canada to southern Mexico

Habitat: A variety of habitats, including deserts, swamps, and forests

Diet: Many types of small mammals such as rats, mice, hares, rabbits, opossums, fawns, and sometimes birds and reptiles

FUN FACT

Bobcats are about twice the size of an average house cat.

TIGER *(PANTHERA TIGRIS)*

Tigers are the largest of all cat species. They are easily identified by their dark stripes and their orange-and-white coats. Some tigers are also all white with dark stripes. Tigers are burly and powerful, and they are able to take down prey several times heavier than they are. Their wide paws and long claws help them attack prey, and then they deliver a deadly bite.

HOW TO SPOT

Size: 4.8 to 9.5 feet (1.5 to 2.9 m) long; tail length is about 2.3 to 3.6 feet (0.7 to 1.1 meters); 165 to 716 pounds (75 to 325 kg)

Range: Asia, India, and Russia

Habitat: Mangrove forests, arid forests, tropical forests, and northern-latitude coniferous forests

Diet: Mostly hoofed mammals such as pigs and deer, but sometimes bears, rhinos, and young elephants

FUN FACT

There are different tiger subspecies around today. They are the Amur, Bengal, Indochinese, Malayan, South China, and Sumatran tigers.

AFRICAN LION *(PANTHERA LEO)*

Lions are large, carnivorous predators at the top of the food chain in their habitats. They are the only species of cat that lives in groups. A group of lions is called a pride. They have yellow-gold coats. Male lions have a mane. Their manes show off their fitness to potential mates and are used to size up rival males.

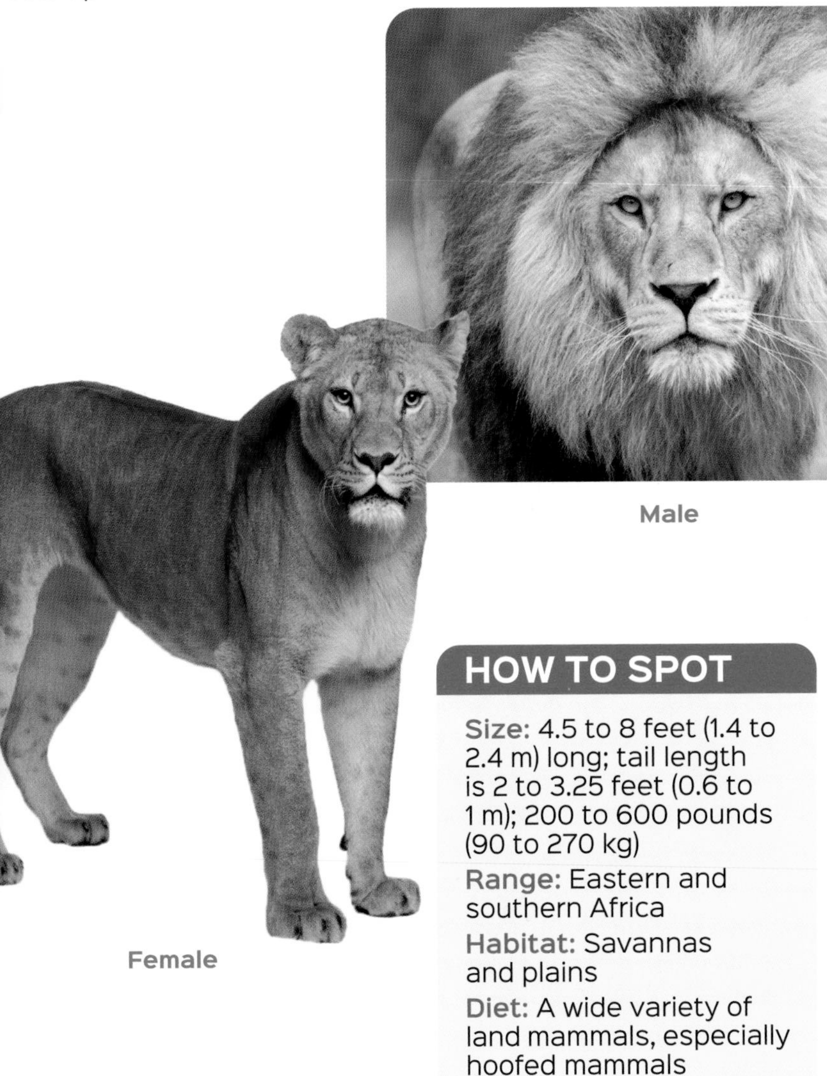

Male

Female

HOW TO SPOT

Size: 4.5 to 8 feet (1.4 to 2.4 m) long; tail length is 2 to 3.25 feet (0.6 to 1 m); 200 to 600 pounds (90 to 270 kg)

Range: Eastern and southern Africa

Habitat: Savannas and plains

Diet: A wide variety of land mammals, especially hoofed mammals

CHEETAH *(ACINONYX JUBATUS)*

Cheetahs have solid-black spots, a slim body, and long, slender legs. Their bodies are built for speed. They also have hard footpads that give them extra traction and claws that work like cleats. In addition, cheetahs use their long tails to stabilize their bodies when sprinting and can use their tails to help make sharp, sudden turns.

HOW TO SPOT

Size: 3.3 to 5 feet (1 to 1.5 m) long; tail length is 2 to 2.7 feet (0.6 to 0.8 m); 75 to 125 pounds (34 to 57 kg)

Range: Parts of Africa and Iran

Habitat: Grasslands and open ranges

Diet: Gazelles, antelope, hares, birds, and rodents

Cheetahs have tear marks on their faces.

FUN FACT

Cheetahs are the fastest land mammal. They are able to reach speeds of 70 miles per hour (112 kmh) when sprinting.

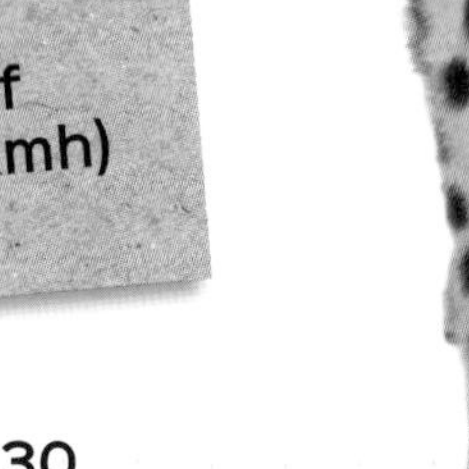

SNOW LEOPARD *(PANTHERA UNCIA)*

Snow leopards live in high mountain habitats. Their light colorings with gray-and-black markings allow them to blend in with their rocky surroundings. They are built for climbing. Their long, thick tails help them balance on dangerous rock cliffs. Their tails are also built-in scarfs. A snow leopard will wrap itself in its tail for extra warmth.

DOMESTIC CATS

Thousands of years ago, humans began to settle in communities and store food. This attracted rodents. The first wildcats to enter towns must have loved the abundance of prey, while humans loved the pest control. Eventually, some wildcats became domesticated.

HOW TO SPOT

Size: 2.8 to 4 feet (0.9 to 1.2 m) long; tail length is 2.75 to 3.4 feet (0.8 to 1 m); 48 to 121 pounds (22 to 55 kg)

Range: High mountains of central Asia

Habitat: High, steep, rugged mountains

Diet: Blue sheep, ibex, marmots, game birds, and small rodents

GRAY WOLF *(CANIS LUPUS)*

The coats of gray wolves are often gray, but they can also be black, reddish, brown, or white. They are the largest of all canines and are known for their distinctive howls, which are how they communicate with one another. They are highly social, living in packs of two to 15 individuals, usually family members. Each pack has a strict social order, works together to raise the pups, and coordinates hunts.

HOW TO SPOT

Size: 28 to 31 inches (71 to 79 cm) at the shoulder; tail length is 13 to 20 inches (33 to 51 cm); 40 to 175 pounds (18 to 80 kg)

Range: Alaska, western and northeastern Canada, northwestern United States, and certain areas of Europe and Asia

Habitat: Forests and tundra

Diet: Large mammals, including elk, deer, moose, and caribou, as well as hares and beavers

FUN FACT

Gray wolves that live near the coast in Canada will swim between islands and will eat seafood, including crabs and clams.

DOMESTIC DOGS

Dogs were the first animal to be domesticated by humans. Humans used dogs to help with hunting and herding, as guards, and for companionship. Today, there are hundreds of domestic dog breeds, but they all share a common ancestor: the gray wolf.

ARCTIC FOX *(VULPES LAGOPUS)*

Arctic foxes are small canines with white fur in the winter and gray-brown fur in the summer. This allows them to blend into the habitat as the seasons change. Other adaptations that help them survive the Arctic cold are compact bodies, small ears, and thick fur. They also create dens with a system of tunnels that are used generation after generation and can have dozens of entrances.

HOW TO SPOT

Size: 12 inches (30 cm) at the shoulder; tail length is up to 13.75 inches (34.9 cm); 6.5 to 17 pounds (2.9 to 7.7 kg)

Range: Arctic regions of Europe, Asia, North America, Greenland, and Iceland

Habitat: Arctic tundra, generally near coastlines

Diet: Small rodents, Arctic hares, and birds

AFRICAN WILD DOG *(LYCAON PICTUS)*

African wild dogs have mottled fur with patches of brown, yellow, black, and white. They are highly social dogs and live in packs averaging seven to 15 members. Within the pack there is a social structure, including a dominant male and dominant female. They communicate with one another through vocalizations, touch, and actions. The whole pack takes care of the young, the old, and the sick or injured.

HOW TO SPOT

Size: 26 to 30 inches (65 to 75 cm) at the shoulder; tail length is 12 to 16 inches (30 to 40 cm); 40 to 79 pounds (18 to 36 kg)

Range: Sub-Saharan Africa

Habitat: Savannas, grasslands, and open woodlands

Diet: Other mammals, including both small prey and prey twice the wild dog's size, such as antelope and zebras

Large, round ears help African wild dogs hear prey.

FUN FACT

After a hunt, African wild dogs will regurgitate, or spit back up, the meat they've eaten for the pups, wounded, or others that did not participate in the hunt.

DINGO *(CANIS LUPUS DINGO)*

Dingoes have golden or reddish fur. They can be found in packs of up to 12 members or alone. Their hunting behaviors are similar. Sometimes they hunt alone, while other times they hunt as a coordinated pack. While dingoes are not a threatened species, scientists believe the purebred species may become threatened because they breed with domestic dogs.

HOW TO SPOT

Size: 18.5 to 26 inches (47 to 66 cm) at the shoulder; tail length is 12 to 13 inches (30 to 33 cm); 22 to 33 pounds (10 to 15 kg)

Range: Australia and Southeast Asia

Habitat: Forests, plains, mountains, and deserts (if watering holes are available)

Diet: Small animals, including rabbits, rodents, birds, and lizards. They will also eat some fruit and plants.

CANINES

Canines are found all over Earth, except in Antarctica. Most are thin with long legs and a long muzzle. Canines are considered carnivores, but most are actually omnivores. Animals they kill themselves, carrion, and plants are all food sources. When they hunt prey, many canines do so in packs.

DESERT KANGAROO RAT

(DIPODOMYS DESERTI)

Desert kangaroo rats can be gray to pale brown, have a long tail, and have large hind legs. Like the kangaroos they share a name with, these rats move by hopping, using their long tails for stability. Kangaroo rats dig deep burrows in sandy soil. These burrows have a complex network of tunnels and entrances, as well as areas for food storage and a nesting room.

HOW TO SPOT

Size: 4.1 to 6.3 inches (10.4 to 16 cm) long; tail length is 7.3 to 8.3 inches (18.5 to 21 cm); 2.6 to 5.2 ounces (73 to 148 g)

Range: Southwestern North America

Habitat: Hot deserts and sand dunes

Diet: Dried plant materials and creosote bush seeds

The desert kangaroo rat has a white line across its back.

FUN FACT

Kangaroo rats can go long periods of time without drinking water. Their bodies are adapted to extract wat from the food they eat.

RODENTS

There are more than 2,000 different species of rodents. The common trait among them is their teeth. All rodents have teeth specially adapted to gnawing. They each have a pair of pointy teeth on the upper and lower jaw, followed by a gap, and then molars. Their incisors grow continually throughout their lives.

EASTERN GRAY SQUIRREL

(SCIURUS CAROLINENSIS)

The eastern gray squirrel is gray on top with lighter, cream-colored underparts. It also has a long, bushy tail. Eastern gray squirrels do not hibernate. Instead, in the fall they store food to prepare for the winter. This food is buried in various places. Squirrels use both memory and smell to find the food again when needed.

HOW TO SPOT

Size: 9 to 11.25 inches (22 to 28.5 cm) long; tail length is 6.25 to 10.25 inches (15.9 to 26 cm); 0.75 to 1.5 pounds (0.3 to 0.7 kg)

Range: Across the United States and parts of southern Canada

Habitat: Hardwood forests; urban and suburban areas with trees

Diet: Nuts, insects, acorns, seeds, fungi, fruit, and flowers

FUN FACT

There are more than 200 species of squirrels in the world. They range in size from the African pygmy squirrel that is only 5 inches (13 cm) long from nose to tail, to the Indian giant squirrel that measures 3 feet (0.9 m) long.

MUSKRAT *(ONDATRA ZIBETHICUS)*

Muskrats spend their lives in or near water and are well adapted for this lifestyle. They are excellent swimmers and divers, sometimes staying underwater for more than ten minutes. In the winter, they will even dive underneath the surface ice in order to find food. Muskrats have small eyes and ears, slightly webbed feet, and a scaly, flat tail. They also have dark-brown backs that are lighter on the sides.

HOW TO SPOT

Size: 10.2 to 14.2 inches (26 to 36 cm) long; tail length is 7.5 to 10.6 inches (19 to 27 cm); 1.5 to 3.3 pounds (0.7 to 1.5 kg)

Range: Across the United States and Canada

Habitat: Wetlands, including marshes, swamps, and ponds

Diet: Mostly green vegetation, including aquatic plants

IS IT A BEAVER OR A MUSKRAT?

There are some major differences between beavers and muskrats. Beavers are much heavier. Also, the lodges of beavers are bigger than those of muskrats and are made from different materials. Beavers have lodges made from sticks and branches. Muskrats use plant material such as cattails.

AMERICAN BEAVER

(CASTOR CANADENSIS)

Beavers are the largest rodents in North America. They have dark-brown fur; a wide, flat tail; and long, sharp, orange front teeth. They create watertight dams and lodges with branches, mud, and sticks. These activities greatly alter the environment, creating important habitats for many plants and animals. It is for this reason that beavers are called ecosystem engineers.

HOW TO SPOT

Size: 2 to 3 feet (60 to 90 cm); tail length is 11.8 to 15.7 inches (30 to 40 cm); 35 to 70 pounds (16 to 32 kg)

Range: Across the United States and Canada

Habitat: Lakes, rivers, streams, and swamps in woodland areas

Diet: Aquatic plants, leaves, twigs, and roots; bark in the winter

FUN FACT

Beaver lodges usually have their entrances underwater. They also have thick, insulated walls, bedding made from wood shavings, and a "chimney" for ventilation.

YELLOW-BELLIED MARMOT

(MARMOTA FLAVIVENTRIS)

Yellow-bellied marmots are stocky, yellow-brown rodents that are very vocal. Their different whistles are either used to alert others or are used as a threat. Teeth chattering is used as a threat too. Marmots will also scream out of fear or excitement. During warmer months they spend much of the time sunning themselves on rocks, but they hibernate during the winter. Hibernation can last from September to May in some areas.

HOW TO SPOT

Size: 13 to 18.9 inches (33 to 48 cm) long; tail length is 5.1 to 8.7 inches (13 to 22 cm); 3 to 11.5 pounds (1.4 to 5.2 kg)

Range: Mountains across the western United States and Canada

Habitat: A variety, including alpine zones, woodlands, and semideserts

Diet: A wide variety of plants, including grasses and flowers

BRAZILIAN PORCUPINE

(COENDOU PREHENSILIS)

Brazilian porcupines spend most of their lives in trees and have a tail that helps them grab branches. Like the more than 20 other porcupine species, Brazilian porcupines are covered in quills that are their main defense against predators. These quills are made of keratin, the same material as human fingernails. The quills are sharp and keep most predators away.

FUN FACT

A group of porcupines is called a prickle.

HOW TO SPOT

Size: 12 to 24 inches (30 to 60 cm) long; tail length is 13 to 19 inches (33 to 48 cm); 4 to 11 pounds (1.8 to 5 kg)

Range: Northern and central South America

Habitat: Rain forests; forests

Diet: Flowers, leaves, roots, shoots, stems, and buds

NAKED MOLE RAT

(HETEROCEPHALUS GLABER)

Naked mole rats are not entirely naked. Their wrinkly, pink skin does have a few fine hairs that work like whiskers to help them feel around in their dark surroundings. They spend their entire lives underground in a complex network of tunnels. The network includes an area for food storage, a nursery, and even a place to get rid of body waste. A colony may have more than 300 members, led by a queen. Each member has a job to do to contribute to the colony.

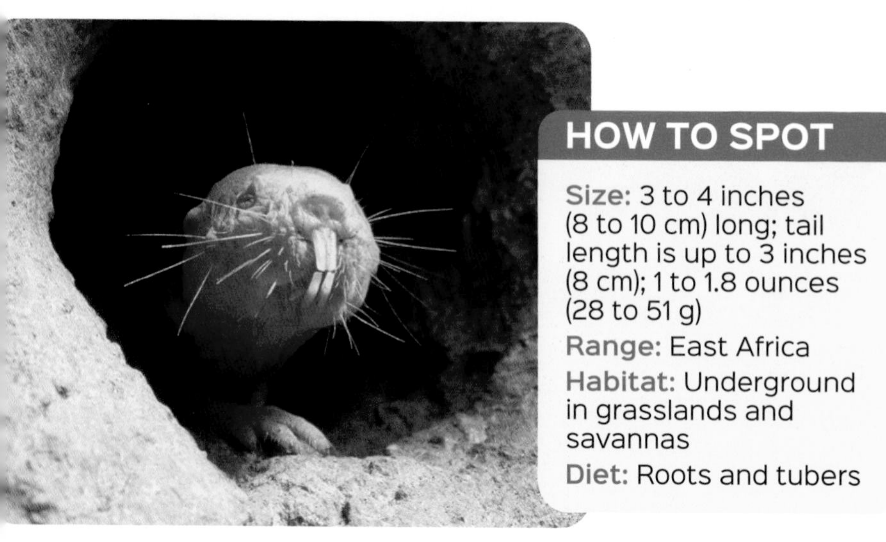

HOW TO SPOT

Size: 3 to 4 inches (8 to 10 cm) long; tail length is up to 3 inches (8 cm); 1 to 1.8 ounces (28 to 51 g)

Range: East Africa

Habitat: Underground in grasslands and savannas

Diet: Roots and tubers

CAPYBARA

(HYDROCHOERUS HYDROCHAERIS)

Capybaras are the largest rodents in the world. They have shaggy hair that is light brown in color, and they are always near water. Their webbed feet make them excellent swimmers. Their nostrils, eyes, and ears are located on the top of their heads, allowing them to lay low in the water to watch for danger.

FUN FACT

To better digest the tough plants capybaras eat, these creatures regurgitate their food and chew it again.

HOW TO SPOT

Size: 3.2 to 4.2 feet (0.9 to 1.3 m); 60 to 174 pounds (27 to 79 kg)

Range: Northern and central South America

Habitat: Flooded grasslands, marshes, low-lying woodlands, and riverbanks

Diet: Aquatic plants, grasses, grains, and reeds

EASTERN COTTONTAIL RABBIT

(SYLVILAGUS FLORIDANUS)

There are many types of cottontail rabbits, and all are named for their puffy, white cotton-ball tail. They have tall ears, but the length depends on location. Those in the warmer, southern regions have longer ears than those in the northern part of their ranges.

HOW TO SPOT

Size: 12.2 to 15.7 inches (31 to 40 cm); tail length is 1.6 to 3.1 inches (4 to 8 cm); 1.75 to 3.4 pounds (0.8 to 1.5 kg)

Range: Southeastern Canada, northwestern South America, and across the United States from the east to the Great Plains

Habitat: Edges of open spaces, including meadows, prairies, and farms

Diet: Grasses and other green plants in the summer; bark, twigs, and buds in the winter

FUN FACT

When cottontails run from a predator, they dart away in a zigzag pattern. They can reach speeds of up to 18 miles per hour (29 kmh).

AMERICAN PIKA *(OCHOTONA PRINCEPS)*

The squeaks and calls of American pikas echo across their mountainous environments. They communicate like this to warn of predators and during mating season. Pairs have even been observed calling together in a duet. American pikas have short ears and limbs. Their fur can change color with the seasons and can be a mixture of off-white in some areas and gray or cinnamon brown in others.

HOW TO SPOT

Size: 7 to 8 inches (18 to 20 cm); tail length is 0.5 inches (1.3 cm); 4.2 to 6.2 ounces (119 to 176 g)

Range: Mountainous regions in southwestern Canada and western United States

Habitat: Boulder fields near alpine meadows, generally above the tree line

Diet: They are herbivores, eating mostly nongrass plants and shrubs

RABBITS, HARES, AND PIKAS

Rabbits, hares, and pikas are all related. They are not rodents. They belong to a separate group called lagomorphs. Rabbits, hares, and pikas are small- to medium-sized land mammals. They are herbivores and have long, sharp teeth that grow throughout their lives to help them chew plants and bark.

ARCTIC HARE *(LEPUS ARCTICUS)*

Arctic hares live in the harsh Arctic climate and are well adapted to it. They have a thick, white coat during the snowy months that provides warmth and camouflage. However, the tips of their ears are black. In southern parts of their ranges, their fur turns gray brown in the summer to match the color of the rocks and vegetation.

FUN FACT

Even if their mothers are mostly white all the time, young Arctic hares are born brown.

HOW TO SPOT

Size: 18.5 to 22.8 inches (47 to 58 cm); tail length is 3 inches (7.6 cm); 5.5 to 13 pounds (2.5 to 6 kg)

Range: Northern Canada, Greenland, and Arctic islands

Habitat: Tundra

Diet: Green plants and berries in the summer; woody plants, mosses, and lichen in the winter

SNOWSHOE HARE

(LEPUS AMERICANUS)

Snowshoe hares are named for their large, furry feet that act as snowshoes. Their feet help them move easily across the top of deep snow. In the winter, their fur is the color of the snow. Their coats change to brown as spring arrives, keeping them well camouflaged. They are agile and quick, with an excellent sense of hearing. Both adaptations help them evade predators.

HOW TO SPOT

Size: 13.8 to 18.5 inches (35 to 47 cm); tail length is 1.2 to 2 inches (3 to 5 cm); 2 to 4 pounds (0.9 to 1.8 kg)

Range: Across Canada, Alaska, and the northern United States; in mountainous regions in the United States

Habitat: Forests with dense undergrowth

Diet: Grasses, herbs, flowers, and berries

FUN FACT

Hares will take dust baths as a way to rid themselves of parasites.

NORTHERN RIVER OTTER

(LONTRA CANADENSIS)

Northern river otters are semiaquatic mammals with short legs, a thick tail, and wide, webbed feet. They have dark-brown fur, with round heads and little ears. These otters are excellent swimmers and divers. On land they can also travel long distances and can run quite fast. They are playful creatures, often sliding down icy or muddy slopes.

HOW TO SPOT

Size: 2.1 to 2.5 feet (0.6 to 0.7 m); tail length is 1 to 1.7 feet (0.3 to 0.5 m); 10 to 24 pounds (4.5 to 11 kg)

Range: Across Canada and Alaska; eastern and northwestern United States

Habitat: Both freshwater and coastal habitats, including rivers, streams, swamps, lakes, and coastlines

Diet: Mainly aquatic organisms such as amphibians, fish, turtles, crayfish, and shellfish

SEA OTTER *(ENHYDRA LUTRIS)*

A sea otter's fur is brown and very thick. These marine mammals spend most of their lives in the ocean, even giving birth and nursing there. To rest, sea otters float on their backs while in the water. They dive for food, then return to floating on their backs to eat. They've also been observed using rocks as tools to open shells.

FUN FACT

Sea otters have the thickest fur of all mammals—up to one million hairs per square inch.

HOW TO SPOT

Size: 2.4 to 3.6 feet (0.4 to 1.1 cm); tail length is 10 to 14 inches (25 to 36 cm); 35 to 100 pounds (16 to 45 kg)

Range: Coastal waters of the North Pacific, including Japan, Alaska, Canada, and California

Habitat: Rocky shores and kelp forests

Diet: Sea urchins, lobsters, abalones, clams, and mussels

Sea otter pup

AMERICAN MINK *(NEOVISON VISON)*

American minks are long, slender mammals with short, stubby legs, partially webbed feet, and a long tail. In fact, the tail makes up to half the mink's total length! Their fur is brown to black in color. It is also thick and covered in oil, making it waterproof. They are excellent swimmers and divers, spending much of their time in the summer hunting for food in the water.

HOW TO SPOT

Size: 12 to 16 inches (30 to 41 cm) long; tail length is 5.75 to 8 inches (15 to 20 cm); 1 to 2.25 pounds (0.45 to 1 kg)

Range: Across Canada, Alaska, and other areas of the United States

Habitat: Forested areas close to streams, marshes, lakes, and ponds

Diet: Small mammals, crayfish, birds, snakes, and frogs

BLACK-FOOTED FERRET

(MUSTELA NIGRIPES)

Like minks and weasels, black-footed ferrets have long, slender bodies, short legs, and a long tail. They also have distinct, black masks on their faces, as well as black feet and a black tip on their tails. Much of their lives are lived underground in abandoned prairie dog tunnels that they use for travel and shelter.

HOW TO SPOT

Size: 18 to 24 inches (46 to 61 cm) long; tail length is 5 to 6 inches (13 to 15 cm); 1.5 to 2.5 pounds (0.7 to 1.1 kg)

Range: Isolated locations in the central United States

Habitat: Prairies that have prairie dog colonies

Diet: Mostly prairie dogs, but also mice, squirrels, rabbits, birds, and other small animals

FUN FACT

Black-footed ferrets sleep almost 21 hours per day. They sleep during daylight hours and come out at night to hunt prairie dogs.

BACK FROM THE BRINK

By the second half of the 1900s, black-footed ferrets were almost extinct. To prevent full extinction, scientists and government agencies captured the last 18 ferrets and began a breeding program. Captive breeding programs have resulted in thousands of ferrets being reintroduced into the wild.

WOLVERINE *(GULO GULO)*

Wolverines have short legs and claws that can partly retract. Their long fur is blackish brown, and they have a light-colored stripe that runs along the side of their bodies. Wolverines are extremely powerful. They eat all kinds of meat and are able to take down prey that is bigger than them, including small bears and deer. In search of food, wolverines may travel 15 miles (24 km) in a day.

HOW TO SPOT

Size: 2 to 3 feet (0.6 to 1 m) long; tail length is 8 to 10 inches (21 to 26 cm); 18 to 44 pounds (8 to 20 kg)

Range: Northern latitudes of North America, Europe, and Asia

Habitat: Remote tundra and northern alpine forests

Diet: Other mammals, roots, berries, and eggs

HONEY BADGER *(MELLIVORA CAPENSIS)*

Honey badgers, also called ratels, belong to the weasel family. They are stocky, strong, and feisty. The honey badger's hair is almost all black, but it has a white-gray stripe along its head, back, and tail. It has long claws and sharp teeth to rip into meat. These badgers will eat just about anything. However, the favorite food of this mammal is honey, which is how it got its name.

FUN FACT
Honey badgers have extremely thick skin. It can withstand bee stings and porcupine quills.

HOW TO SPOT

Size: 2.4 to 3.2 feet (0.7 to 1 m) long; tail length is 5 to 9 inches (13 to 23 cm); 13.6 to 30 pounds (6.2 to 13.6 kg)

Range: Sub-Saharan Africa, western Asia, Iran, and Saudi Arabia

Habitat: Tropical and subtropical forests; grasslands

Diet: Honey and honeybee larvae, insects, amphibians, reptiles, birds, small mammals, roots, bulbs, berries, and fruit

STRIPED SKUNK *(MEPHITIS MEPHITIS)*

Striped skunks are easily recognized by their black fur and the two thick, white stripes that run the length of their bodies and tails. Skunks are also well-known for their defense adaptation. When striped skunks are threatened, they spray a foul-smelling liquid from the base of their tails. Not only is the odor extremely strong, but it can burn the eyes or cause temporary blindness in an attacker.

FUN FACT

At eight days old, baby skunks can spray their foul-smelling liquid. They can do this before they even open their eyes! They open their eyes about 14 days later.

HOW TO SPOT

Size: 12 to 20 inches (30 to 50 cm) long; tail length is 9 to 14 inches (23 to 36 cm); 1.5 to 13 pounds (0.7 to 6 kg)

Range: Northern Mexico to southern Canada

Habitat: A variety of habitats, including open woods, deserts, plains, and urban and suburban areas

Diet: Varies depending on what is available, but includes insects, eggs, fruit, small mammals, reptiles, and amphibians

RACCOON *(PROCYON LOTOR)*

Sometimes called a masked bandit because of the dark mask over their eyes, raccoons are highly adaptable mammals. They thrive in both tropical and cold areas. They also eat a wide variety of both plants and animals, depending on what is available. Their front paws are similar to human hands, allowing them to pick up their food, feel around inside a tree for insects or in water for aquatic prey, and even pull lids off of trash cans.

HOW TO SPOT

Size: 16 to 24 inches (40 to 60 cm) long; tail length is 6 to 16 inches (15 to 40 cm); 5 to 33 pounds (2.3 to 15 kg)

Range: Across the United States, southern Canada, and northern South America

Habitat: Prefers moist woodland areas, but can be found in urban and suburban areas, and wetlands

Diet: Will eat a variety of aquatic species, mice, insects, eggs, plants, and fruit

DWARF MONGOOSE

(HELOGALE PARVULA)

Dwarf mongooses are the smallest of all mongoose species. They have smooth, speckled fur that is brown to black in color. Relative to their sizes, they have long tails. They live in packs of up to 32 animals. Within this group, there will be only one breeding pair. The female in that pair is the dominant animal in the whole pack. Her mate is the second ranked.

HOW TO SPOT

Size: 7 to 10 inches (18 to 25 cm); tail length is 4.7 to 8 inches (12 to 20 cm); 1 pound (0.45 kg)

Range: Across sub-Saharan Africa

Habitat: Woodlands, mountain scrublands, and grasslands

Diet: Mostly insects; also eggs, fruit, and small animals

MEERKAT *(SURICATA SURICATTA)*

Meerkats, a species of mongoose, are highly social and cooperative mammals. They live in groups called mobs that may have up to 30 individuals and several families. Meerkats have long bodies covered in fur that can range from brown to gold, orange, or silver, with a dark mask around their eyes. Meerkats are often seen standing upright on their rear legs, using their tails for support. These individuals are serving as lookouts for predators while others forage for food.

FUN FACT

Meerkats like to bask in the sun in the morning. This helps to warm them up after a chilly night.

HOW TO SPOT

Size: 10 to 14 inches (25 to 36 cm) long; tail length is 7 to 10 inches (18 to 25 cm); 1.4 to 2.1 pounds (0.6 to 0.95 kg)

Range: Areas of southern Africa

Habitat: Arid savannas, grasslands, and plains

Diet: Primarily insects; sometimes small animals, eggs, and some plants

TWO-TOED SLOTH *(MEGALONYCHIDAE)*

These sloths are distinct from their three-toed cousins in that they have only two toes on their front feet and three on the back. They are also slightly larger. They have short heads and necks, and long limbs with curved claws 3 to 4 inches (8 to 10 cm) long. They spend almost their entire lives in the trees, where they sleep, eat, mate, and give birth. They are great swimmers too. When the rain forest floods, sloths will drop from a branch and swim to another tree.

FUN FACT

Sloths only come down to the ground to get rid of body waste. This happens about once a week.

HOW TO SPOT

Size: 21 to 29 inches (53 to 74 cm); 9 to 17 pounds (4 to 8 kg)

Range: Central and South America

Habitat: Tropical rain forest canopies

Diet: Leaves, fruit, stems, buds, and sometimes insects

THREE-TOED SLOTH *(BRADYPODIDAE)*

Like their two-toed cousins, three-toed sloths move very little. However, this is an excellent adaptation. Both the two-toed and three-toed sloths move so little that algae actually grows on their fur. This acts as camouflage in the tropical rain forest. In addition, if they don't move much, it is difficult for predators to spot them. All sloths hang from trees with a strong grip aided by their long claws. The three-toed sloth has three claws on each limb. It also has long yellow-brown or tan fur.

HOW TO SPOT

Size: 23 inches (58 cm); 6.6 to 11 pounds (3 to 5 kg)

Range: Central and South America

Habitat: Tropical rain forest canopies

Diet: Leaves

AARDVARK *(ORYCTEROPUS AFER)*

Aardvarks look to be part pig because of their long snouts. And they look part rabbit because of their long ears. They also look part kangaroo because of their long tails. Yet aardvarks are a separate species. They come out of their cool underground burrows to hunt only at night. Using sharp front claws, aardvarks break into ant or termite mounds. Then they use their long, sticky tongues to feast on the insects inside.

HOW TO SPOT

Size: 3.6 to 4.4 feet (1 to 1.3 m); tail length is 21 to 26 inches (53 to 66 cm); 110 to 180 pounds (50 to 82 kg)

Range: Sub-Saharan Africa

Habitat: Grasslands and savannas

Diet: Ants and termites

GIANT ANTEATER

(MYRMECOPHAGA TRIDACTYLA)

Giant anteaters have a distinct, long snout, powerful front limbs, and sharp claws. They also have an excellent sense of smell so they can determine the species of ants or termites in a mound. These adaptations allow the anteater to break open a mound and slurp up the insects inside. They can flick their extra-long tongues in and out as much as 150 times in a minute. With that tongue, they can eat more than 30,000 insects in one day.

HOW TO SPOT

Size: 3.3 to 4 feet (1 to 1.2 m); tail length is 2.3 to 2.9 feet (0.8 to 1 m); 60 to 140 pounds (27 to 64 kg)

Range: Central and South America

Habitat: Grasslands, wetlands, and forests

Diet: Primarily termites and ants

FUN FACT

Not only is the giant anteater's tongue like a 2-foot-long (0.6 m) strand of spaghetti, but it also has tiny backward-facing spines and is covered in sticky saliva.

EUROPEAN HEDGEHOG

(ERINACEUS EUROPAEUS)

European hedgehogs are small insectivores that got their name from their feeding habits. They forage primarily among hedges—bushes and shrubs growing close together. And when they do this, they grunt like pigs! Their babies are even named hoglets. European hedgehogs are covered in short, sharp spines tipped with white. When threatened, they will curl into a ball to protect their soft underbellies, which also makes them look like an unappetizing ball of spikes.

HOW TO SPOT

Size: 5 to 12 inches (13 to 30 cm); tail length is 1 to 2 inches (2.5 to 5 cm); 0.875 to 2.4 pounds (0.4 to 1.1 kg)

Range: Across Europe and central Asia

Habitat: Temperate climates at the edges of fields or hedgerows; scrublands

Diet: Insects, including ants, bees, earwigs, butterflies, moths, and beetles

STAR-NOSED MOLE

(CONDYLURA CRISTATA)

The star-nosed mole is so named because of the 22 fleshy tentacles that make up its nose and form a star shape. The extremely sensitive tentacles are used to feel for food as star-nosed moles forage through dark tunnels in moist soil. This mammal has dark, dense fur. It also has big claws and wide, scaly feet.

HOW TO SPOT

Size: 6 to 8 inches (15 to 20 cm); tail length is 2.6 to 3.4 inches (6.6 to 8.6 cm); 1.9 ounces (54 g)

Range: Eastern North America

Habitat: Forests with poor drainage, wetlands, ponds, lakes, rivers, and streams

Diet: Insects, some worms, and fish

FUN FACT

Each of the star-nosed mole's tentacles has about 25,000 touch receptors. This allows the mole to sense the electrical currents produced by other animals and to identify prey.

SHORT-EARED ELEPHANT SHREW *(MACROSCELIDES PROBOSCIDEUS)*

Short-eared elephant shrews are the smallest of the 17 species of elephant shrew. They have long, flexible snouts like elephants and are closely related to them. Their snouts are also very sensitive. Short-eared elephant shrews have gray-brown fur, white underparts, and a long tail. For their sizes, short-eared elephant shrews are strong and quick, which helps them evade predators.

HOW TO SPOT

Size: 4 inches (10 cm) long; tail length is 3.8 to 5.1 inches (9.7 to 13 cm); 1 to 1.5 ounces (28 to 43 g)

Range: Southern Africa

Habitat: Deserts, grasslands, and shrublands with soft, sandy soil

Diet: Ants and termites; sometimes berries and shoots of young plants

AMERICAN PYGMY SHREW

(SOREX HOYI)

American pygmy shrews are among some of the world's smallest mammals. They have fur that is different shades of gray. Pygmy shrews have narrow heads, pointed noses with whiskers, and long tails. The tail is generally one-third of a pygmy shrew's total length. Pygmy shrews are well-adapted diggers, burrowing in decomposing leaves or soft, moist soil. They will also use tunnels made by other animals as they search for food.

FUN FACT

When shrews are threatened or upset, they emit a musky liquid from their scent glands. This smell keeps most predators away.

HOW TO SPOT

Size: 2 to 2.8 inches (5 to 7 cm); tail length is 1.1 to 1.77 inches (3 to 4.5 cm); 0.07 to 0.2 ounces (2 to 6 g)

Range: Across Alaska and Canada, northeastern United States, and Colorado's Rocky Mountains

Habitat: Varied, including forests, wetlands, and grasslands

Diet: Ants, spiders, caterpillars, earthworms, beetles, and grubs

SCREAMING HAIRY ARMADILLO

(CHAETOPHRACTUS VELLEROSUS)

Like all armadillos, the screaming hairy armadillo has protective armor. It has a shield on its head, between its ears, and another over its back, called a carapace. The carapace has more than a dozen bands, several of which can move. This allows the armadillo to curl into a ball when threatened by a predator. These armadillos also scream, which is how they got their name.

HOW TO SPOT

Size: 8.7 to 15.7 inches (22 to 40 cm); tail length is 3.5 to 6.9 inches (9 to 17.5 cm); 1.9 pounds (0.9 kg)

Range: Western Bolivia; northwestern Argentina

Habitat: Desert sand dunes

Diet: Plants and insects

FUN FACT

As screaming hairy armadillos forage for food in the sand dunes, they end up eating a lot of sand. Scientists have found some armadillos' stomachs halfway filled with it.

TREE PANGOLIN *(MANIS TRICUSPIS)*

Tree pangolins are covered in scales, which are made of keratin. These scales, along with their long, pointed snouts, have earned them the nickname “scaly anteaters,” even though they are not related to anteaters. All of the scales on the pangolin overlap, creating protective armor. When threatened, they curl into a tight ball that predators cannot uncurl. Mother pangolins will curl around their young, who also roll into a ball.

HOW TO SPOT

Size: 12 to 34 inches (30 to 86 cm); tail length is 11 to 34 inches (28 to 86 cm); 3.7 to 5.3 pounds (1.7 to 2.4 kg)

Range: Central Africa

Habitat: Rain forests

Diet: Insects, mostly termites and ants

BUMBLEBEE BAT

(CRASEONYCTERIS THONGLONGYAI)

The bumblebee bat—also called the Kitti's hog-nosed bat—is one of the smallest mammals on Earth. It can fit in the palm of a person's hand. These mammals do not have tails. But they do have a nose like the snout of a pig, and they have large ears. Their upper bodies may be gray and reddish brown, while their undersides are paler. These bats live in small colonies of approximately 100 bats. They are only active a few minutes every day—about 18 minutes at dawn and 30 minutes at dusk. Like other bats, they hunt using echolocation and catch their prey in mid-flight.

UPSIDE DOWN BATS

Most bats roost upside down. Instead of needing to launch themselves into the air vertically to fly, bats simply let go. When they drop from their roosts, they can start flying right away.

HOW TO SPOT

Size: 1.14 to 1.3 inches (2.9 to 3.3 cm); wingspan is up to 6.7 inches (17 cm); 0.06 to 0.07 ounces (1.7 to 2 g)

Range: Southeast Asia

Habitat: Limestone caves in bamboo forests

Diet: Insects and spiders

MEXICAN FREE-TAILED BAT

(TADARIDA BRASILIENSIS)

These gray-brown, medium-sized bats migrate from the southern United States to spend the winter in the warmer climates of Mexico. Mexican free-tailed bats choose a variety of sites for roosting, including caves, the underside of bridges, and home attics. Roost sites can include millions of bats. Just prior to emerging from the roost site at night, there is considerable squeaking and chattering, followed by a roaring sound as they whoosh out into the night sky.

HOW TO SPOT

Size: 2.2 to 2.6 inches (5.6 to 6.6 cm); wingspan is 11 inches (28 cm); tail length is 1.1 to 1.7 inches (2.8 to 4.3 cm); 0.35 to 0.49 ounces (10 to 14 g)

Range: Southern United States, Mexico, Central America, and northern South America

Habitat: Caves, deserts, scrublands, and urban and suburban areas

Diet: Insects, including moths and bees

FUN FACT

There are approximately 1,240 species of bats. They are found everywhere except in the colder Arctic and Antarctic regions.

JAMAICAN FRUIT BAT

(ARTIBEUS JAMAICENSIS)

The Jamaican fruit bat is also known as the Mexican fruit bat. This mammal has a distinct, leaf-shaped nose that sticks out on its muzzle, and it doesn't have a tail. Its short fur is black, brown, or gray. This bat also has four white stripes on its face. One stripe is below and above each of its eyes.

FUN FACT

While feeding at night, between a dozen and several hundred Jamaican fruit bats have been seen gathering at one fruit tree.

HOW TO SPOT

Size: 2.75 to 3.34 inches (7 to 8.5 cm); wingspan is 1.8 to 2.6 inches (4.6 to 6.6 cm); 1.4 to 2.1 ounces (40 to 60 g)

Range: South and Central America

Habitat: Rain forests, dry forests, and plantations

Diet: Fruit such as figs, guavas, bananas, and papayas

HONDURAN WHITE BAT

(ECTOPHYLLA ALBA)

Honduran white bats have fur that ranges from white to gray. Their inner wings are dark in color, and their outer wings are lighter. These bats also have an amber or yellow nose shaped like a leaf, and they don't have a tail. Honduran white bats live in groups and roost in large leaves. These bats are not very territorial. Animals such as snakes, owls, raptors, squirrel monkeys, and opossums will prey on them.

HOW TO SPOT

Size: 1.45 to 1.85 inches (3.7 to 4.7 cm); wingspan is 4 inches (10.2 cm); 0.2 ounces (5.7 g)

Range: Honduras and other Central American countries

Habitat: Rain forests

Diet: Figs and other fruit

INDIAN FLYING FOX

(PTEROPUS GIGANTEUS)

The Indian flying fox is a large bat species, and it has a coat that's reddish brown. It also has large eyes, small ears, and a long snout. Hundreds of these bats will roost in a single tree, and they will travel up to around 9 miles (15 km) to search for food.

FUN FACT

The Indian flying fox learns how to fly when it is 11 weeks old.

HOW TO SPOT

Size: 9 inches (23 cm); wingspan is 3.9 to 4.9 feet (1.2 to 1.5 m); 21.2 to 56.4 ounces (600 to 1,600 g)

Range: South Central Asia

Habitat: Swamps and forests

Diet: Fruit such as mangoes, figs, and guavas

VAMPIRE BAT *(DESMODUS ROTUNDUS)*

Vampire bats have short fur that ranges from reddish orange to brown. These are one of the only bat species that drink blood. When they've approached their prey, they use their sharp teeth to cut open the prey's skin. Then these bats lick up the blood coming from the incision. Vampire bats are also the only bat that can launch from the ground, so they can fly to the ground to capture prey by crawling or hopping to it.

HOW TO SPOT

Size: 3.5 inches (8.9 cm); wingspan is 7 inches (17.8 cm); 2 ounces (57 g), but can range depending on how much blood is consumed

Range: Mexico; Central and South America

Habitat: Tropical areas

Diet: Blood

GRAY MOUSE LEMUR

(MICROCEBUS MURINUS)

Gray mouse lemurs are some of the smallest primates in the world. They have brownish-gray fur, short arms and legs, and a long tail. They spend most of their lives in the trees, jumping from one tree to the next to move around. They forage at night, and during the day, females rest in groups of up to 15. Males sleep in pairs away from the females.

HOW TO SPOT

Size: 4.7 to 5.5 inches (12 to 14 cm); tail length is 5.1 to 5.7 inches (13 to 14.5 cm); 2.1 ounces (60 g)

Range: Madagascar

Habitat: Tropical and scrub forests

Diet: Mostly insects, sometimes plants, fruit, flowers, and even small reptiles

WHAT IS A PRIMATE?

The group of mammals known as primates includes monkeys, apes, lemurs, and humans. Primates possess a high level of intelligence and large brain size. Some use objects in their environments as tools. In addition, primates have eyes that face forward and good eyesight.

RING-TAILED LEMUR *(LEMUR CATTA)*

Ring-tailed lemurs have alternating black and white rings on their tails. Their backs and limbs are gray, and their bellies are white. Unlike many other lemurs, ring-tailed lemurs spend time on the ground, moving around on all fours. They live in troops of up to 30, including both males and females. The females live with the troop they were born into, while males tend to find a new troop.

HOW TO SPOT

Size: 18 inches (46 cm); tail length is up to 24 inches (61 cm); 5 to 7.5 pounds (2.3 to 3.4 kg)

Range: Southwestern Madagascar

Habitat: Arid regions, both open and forested

Diet: Leaves, flowers, and insects

FUN FACT

More than 100 species of lemur exist. All of them are found on the island of Madagascar and on nearby islands off the east coast of Africa.

STINK FIGHTS

Male ring-tailed lemurs have scent glands on both their shoulders and their wrists. The glands produce a foul odor. When there is a fight looming, lemurs will pull their tails across these glands, making their tails a smelly weapon. Then two rivals will wave and flick their tails at each other. The lemur who can stand the smell the longest is the winner.

CENTRAL AMERICAN SPIDER MONKEY *(ATELES GEOFFROYI)*

Central American spider monkeys have very long arms, legs, and tails compared to each animal's individual size. They use their tails for gripping and holding things. They also use their tails to help them move through the trees. These mammals have black hands and feet, and their fur may be black, red, or brown. Their faces often have lighter skin. Spider monkeys spend most of their time in the trees in the rain forest canopy. They live in troops of several dozen individuals.

HOW TO SPOT

Size: 12 to 24.8 inches (30.5 to 63 cm) long; tail length is 25 to 33 inches (63.5 to 84 cm); 16 pounds (7.3 kg)

Range: Mexico, Central America, and northwest South America

Habitat: Tropical rain forests

Diet: Mostly fruit; also nuts, bird eggs, seeds, and insects

BLACK HOWLER MONKEY

(ALOUATTA CARAYA)

Howler monkeys are named for their ability to howl. This sound, like a bark, can be heard up to 3 miles (4.8 km) away. In fact, they are the loudest land animal in the western hemisphere. Males and females have different-colored coats. Females' coats are blonde, and males' coats are black. Howler monkeys also have black faces.

Male

FUN FACT

When black howler monkeys wake up in the morning, the males begin a chorus of howls. This helps establish and defend their territories.

Female

HOW TO SPOT

Size: 20 to 26 inches (51 to 66 cm); tail length is 24 to 26 inches (61 to 66 cm); 16 to 32 pounds (7.3 to 15 kg)

Range: Central South America

Habitat: Arid deciduous forests and rain forests

Diet: Fruit, leaves, and flowers

GUINEA BABOON *(PAPIO PAPIO)*

The Guinea baboon is the smallest species of baboon. They have all of the qualities that distinguish baboons from other monkeys. Their faces and their behinds are hairless. In addition, the males grow a mane when they mature. Guinea baboons are reddish brown, but color varies depending on region. They spend most of the day on the ground but sleep in large trees at night.

HOW TO SPOT

Size: 20 to 45 inches (51 to 114 cm); tail length is 17.9 to 28 inches (45.6 to 71.1 cm); 28 to 57 pounds (13 to 26 kg)

Range: Western equatorial Africa

Habitat: Arid woodlands, savannas, and grasslands

Diet: Seeds, shoots, roots, fruit, and fungi, but also invertebrates and small vertebrates when available

MANDRILL *(MANDRILLUS SPHINX)*

Mandrills can be easily identified by their colorful faces. The thick ridges on the mandrill's nose are blue and purple. Their noses and lips are bright red. These same colors appear on their behinds. The colors will brighten if the mandrill gets excited. Males tend to have brighter colors than females. The coloration is to attract a mate and to help them find one another in the forest.

HOW TO SPOT

Size: 22 to 32 inches (56 to 81 cm); 24 to 73 pounds (11 to 33 kg); females are about half the size of males

Range: Equatorial Africa

Habitat: Rain forests

Diet: Seeds, nuts, fruit, fungi, insects, and small animals

Male

Female

FUN FACT

Mandrills stuff their large cheek pouches with food to eat later or to eat in a safer place.

CHIMPANZEE *(PAN TROGLODYTES)*

Chimpanzees are highly intelligent and are able to make and use tools. They have also been observed seeking out medicinal plants when they are injured or sick. In addition, they use a variety of methods to communicate. This includes body language, gestures, facial expressions, pounding the ground, foot stomping, and a wide range of vocalizations. Chimpanzees have brown to black fur, black or brown faces, and short thumbs and long fingers.

HOW TO SPOT

Size: 4 to 5.5 feet (1.2 m to 1.7 m) tall; 70 to 120 pounds (32 to 54 kg)

Range: Central Africa

Habitat: Tropical rain forests, sometimes mountain forests and savannas

Diet: A variety, including seeds, fruit, leaves, honey, flowers, and insects, as well as small or medium-sized animals

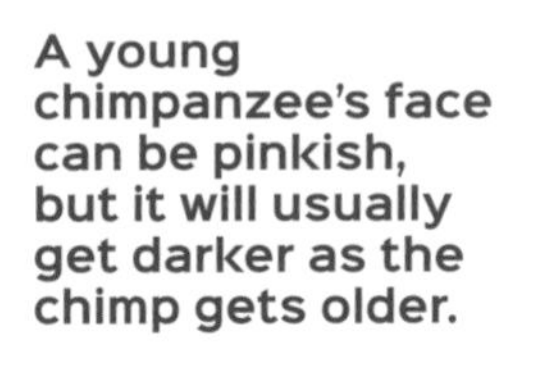

A young chimpanzee's face can be pinkish, but it will usually get darker as the chimp gets older.

BONOBO *(PAN PANISCUS)*

Bonobos were once thought to be a type of chimpanzee. However, bonobos are thinner than chimpanzees, with a smaller head and ears. They teach their young to use tools to find food, share food among one another, and work together for the good of the troop. They also have a wide range of vocalizations, including a call that tells the others where there's a good food source.

FUN FACT
Chimpanzees, bonobos, and other great apes build nests! Each evening, they build sturdy nests in the forks of trees using twigs and leaves.

HOW TO SPOT

Size: 3.6 to 3.9 feet (1.1 to 1.2 m) tall; 68 to 86 pounds (31 to 39 kg)

Range: Central Africa in the Democratic Republic of the Congo

Habitat: Rain forests

Diet: Fruit, seeds, nuts, leaves, flowers, worms, insects, and small fish

MONKEY OR APE?

While they are both primates, apes and monkeys are not the same thing. To tell apes apart from monkeys, start with the tail. Apes do not have tails, while most monkeys do. Most apes are larger than monkeys. Finally, the noses of apes are wider and shorter than those of monkeys.

WESTERN LOWLAND GORILLA

(GORILLA GORILLA GORILLA)

Gorillas are the largest of all the apes. The western lowland gorilla is the smallest gorilla subspecies. They have short, dark fur covering their entire bodies, except their faces. Western lowland gorillas live in groups of five, on average. They are ruled by a dominant male. These males are called silverbacks because of their graying hair.

HOW TO SPOT

Size: 4 to 6 feet (1.2 to 1.8 m) tall; 150 to 440 pounds (68 to 200 kg); adult females are usually about half the size of adult males

Range: Western central Africa

Habitat: Tropical rain forests

Diet: Mostly fruit, shoots, bark, and nuts; sometimes insects or lizards

SUMATRAN ORANGUTAN

(PONGO ABELII)

Sumatran orangutans have long, red fur. Both males and females have beards on their wide, flat faces. These primates have extremely long arms, which helps them move easily through the trees, where they spend almost all of their time. Most nights, Sumatran orangutans create new nests made of sticks, bent branches, and leaves.

FUN FACT

Some primates walk on the ground using all four limbs. This is known as knuckle walking. They use the knuckles on their hands and the soles of their feet while walking.

HOW TO SPOT

Size: 4.3 to 6 feet (1.3 to 1.8 m) tall; 66 to 200 pounds (30 to 90 kg)

Range: Sumatra, an island in Indonesia

Habitat: Rain forests

Diet: Depends on the season, but they will eat figs and other fruit, as well as leaves, bark, flowers, and insects.

AMERICAN BISON *(BISON BISON)*

American bison are the largest land animal in North America. They have a massive head and a notable hump between their shoulders. Both adult males and females have two pointed horns. They are covered in dark fur, with the fur on the head and shoulders thicker and shaggier than the fur on their back halves. When in a blizzard, bison will stand facing the wind, because their heads are the most insulated.

A CASE OF MISTAKEN IDENTITY

Bison and buffalo are two different species, though they are part of the same family. Buffalo are native to Africa and Asia. The American bison is native to North America. However, when European settlers arrived in North America, the animals they encountered looked like buffalo. So that's what they were incorrectly called.

HOW TO SPOT

Size: 5.5 to 6.5 feet (1.7 to 2 m) tall at the hump; 1,800 to 2,400 pounds (816 to 1,089 kg)

Range: Isolated small herds in the American and Canadian West

Habitat: Grasslands and prairies

Diet: Mostly grasses, some flowers, lichen, and leaves

MOUNTAIN GOAT

(OREAMNOS AMERICANUS)

Mountain goats have white fur covering their bodies that gets thicker in the winter. Both males and females have two black horns. These goats have hooves with two toes that spread wide to help with balance. The hooves also have rough pads on the bottom, which help the goats grip the rocks as they move across steep slopes and boulder fields. They are very sure-footed and nimble, able to jump 12 feet (3.6 m) in a single leap.

HOW TO SPOT

Size: 3 to 4 feet (0.9 to 1.2 m) at the shoulder; 100 to 300 pounds (45 to 136 kg)

Range: Northern Rocky Mountains, from Colorado to Alaska

Habitat: Steep, rocky mountains

Diet: Mountain vegetation including plants, grasses, and mosses

FUN FACT

Male goats are called billies. Females are called nannies. And young goats are called kids.

THOMSON'S GAZELLE

(EUDORCAS THOMSONII)

Thomson's gazelles are grazers. Males and females both have horns that grow continuously, and they have hooves split down the middle with two toes. They live in herds of a few gazelles to dozens. They are often found following herds of zebra or gnu, which eat the tall, tough layer of grass. This leaves the shorter, more tender shoots for the gazelles.

FUN FACT

Thomson's gazelles are built for speed. They've been recorded running up to 43.5 miles per hour (70 kmh).

HOW TO SPOT

Size: 20 to 43 inches (51 to 109 cm) at the shoulder; 26 to 165 pounds (12 to 75 kg)

Range: East Africa

Habitat: Arid grasslands and savannas

Diet: Short grasses, leaves, twigs, seeds, grains, and nuts

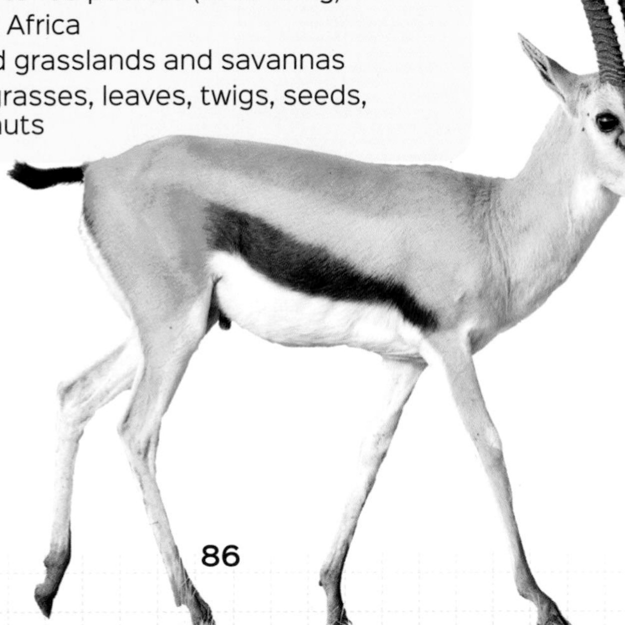

WHITE-TAILED DEER

(ODOCOILEUS VIRGINIANUS)

In the summer, adult white-tailed deer have a reddish-brown coat. In the winter, it becomes a dull gray brown. These animals are lighter on their undersides. The underside of their tails are white too. Young deer, called fawns, have the same color fur but also have white spots, which help them blend in with their habitats. Males, called bucks, shed and then regrow their antlers every year. Bucks use their antlers to fight other bucks over territory or over females.

HOW TO SPOT

Size: 6 to 7.75 feet (1.8 to 2.4 m); 110 to 300 pounds (50 to 136 kg)

Range: Across most of the United States and southern Canada, south into Central America

Habitat: Fields and grasslands; deciduous, coniferous, and mountain forests

Diet: Leaves, grasses, fruit, and nuts

MOOSE *(ALCES ALCES)*

The moose is the largest mammal in the deer family. They are dark brown with a long muzzle. Under each moose's chin is a flap of skin called a bell. Male moose, called bulls, grow huge antlers as wide as 6.5 feet (2 m) across. The antlers have a velvety covering. They shed their antlers and regrow them every year. Moose have poor eyesight but have excellent senses of hearing and smell.

HOW TO SPOT

Size: 6.5 to 7.5 feet (2 to 2.3 m) at the shoulder; 600 to 1,300 pounds (270 to 590 kg)

Range: Across northern North America

Habitat: Wetland areas including swamps, ponds, and willow thickets

Diet: Aquatic plants, twigs, bark, roots, and leaves

A female moose with her young

CARIBOU *(RANGIFER TARANDUS)*

Caribou are covered in hair from the top of their heads all the way down to the bottom of their feet. The hair on their hooves gives them better grip on ice and snow. Both male and female caribou have impressive antlers, especially for their sizes. A female's antlers can be as long as 20 inches (50 cm), and a male's can be as long as 51 inches (130 cm). Males will shed their antlers every November, while females won't shed theirs until calves are born in May. Both will regrow their antlers every year.

HOW TO SPOT

Size: 28 to 53 inches (71 to 135 cm) at the shoulder; 120 to 530 pounds (55 to 240 kg)

Range: Northern North America; Greenland

Habitat: Arctic tundra and boreal forests

Diet: Mosses, ferns, grasses, leaves, and fungi

FUN FACT

Large herds of caribou and reindeer may travel more than 600 miles (970 km) twice a year to access seasonal food sources. This includes plants and grasses on the tundra during summer.

WARTHOG *(PHACOCHOERUS AFRICANUS)*

Warthogs got their name from the bumps on their faces. Yet the bumps are not warts at all, but growths of thick skin. Warthogs have a flat head and four tusks. The upper two tusks are longer than the lower ones, growing to 10 to 11 inches (25 to 28 cm). The lower tusks are about 5 inches (13 cm) long. Warthogs are grazers, using the tusks mostly to fend off predators. If that doesn't work, they are also fast runners, reaching up to 30 miles per hour (48 kmh).

HORNS, ANTLERS, AND TUSKS

Many mammals have horns, antlers, or tusks. But what's the difference? To start, tusks stick out from the mouth even when it's closed. Antlers are made of bone and grow out as part of the animal's skull. Most horns are permanent, growing throughout an animal's life.

HOW TO SPOT

Size: 22 to 33 inches (56 to 84 cm) at the shoulder; 110 to 330 pounds (50 to 150 kg)

Range: Central East Africa

Habitat: Wooded savannas and grasslands

Diet: Grasses, plants, roots, and bulbs

HIPPOPOTAMUS

(HIPPOPOTAMUS AMPHIBIUS)

Hippopotamuses spend their nights grazing for food. They spend their days in the water to stay cool under the hot African sun. With their eyes, ears, and nostrils located on the top of the head, they can see, hear, and breathe while the rest of the body stays underwater.

HOW TO SPOT

Size: 5 to 5.4 feet (1.5 to 1.6 m) at the shoulder; 3,000 to 9,900 pounds (1,400 to 4,500 kg)

Range: Sub-Saharan Africa

Habitat: Always near water, including ponds, lakes, streams, rivers, and swamps

Diet: Primarily grasses

FUN FACT

Hippopotamuses are among Earth's deadliest animals. Not only are they huge in size, but they have a powerful jaw and large teeth.

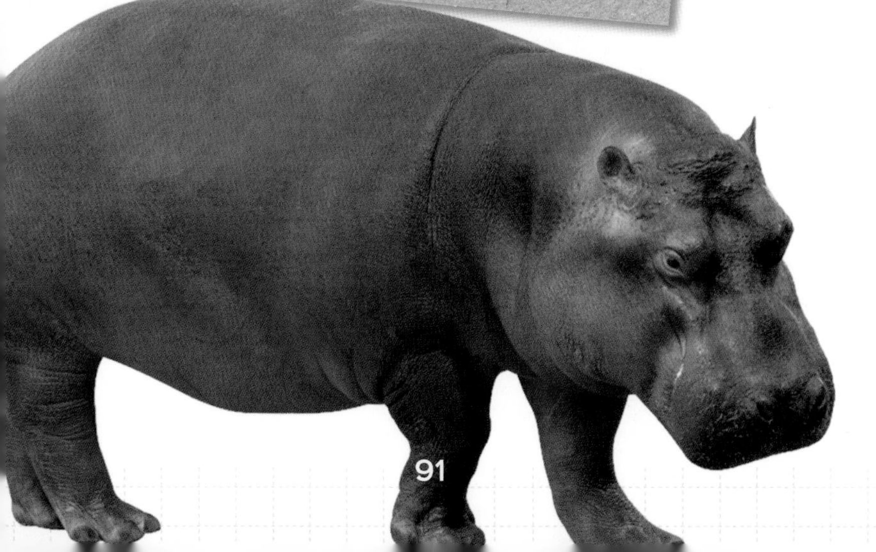

GIRAFFE *(GIRAFFA CAMELOPARDALIS)*

Giraffes are the tallest land mammal. Both their legs and their necks can reach 6 feet (1.8 m) long. These mammals are covered in spots. Each individual has its own unique pattern. Males and females both have ossicones, which are the hair-covered horns on top of their heads. To support such a large body, giraffes have feet approximately 12 inches (30 cm) wide and a heart that weighs 25 pounds (11 kg). Giraffes must eat 16 to 20 hours a day. Luckily, they only need naps of five to 30 minutes a day to recharge.

HOW TO SPOT

Size: 14 to 18 feet (4.3 to 5.5 m) tall; 1,500 to 3,000 pounds (700 to 1,400 kg)

Range: Sub-Saharan Africa

Habitat: Open grasslands and savannas

Diet: Acacia tree leaves and other leaves

DROMEDARY CAMEL

(CAMELUS DROMEDARIUS)

Long-legged dromedary camels are domesticated to carry heavy loads. They have one hump. The humps on camels are not full of water, but fat. The camel can use the fat as an energy source when food and water are scarce. Because of their ability to go long periods of time without food and water in harsh desert climates, camels are often called the ships of the desert.

Camels have two rows of eyelashes, a clear inner eyelid, and the ability to close their nostrils. These things allow them to live in the blowing desert sand.

HOW TO SPOT

Size: 5.9 to 6.6 feet (1.8 to 2 m) at the shoulder; 880 to 1,320 pounds (400 to 600 kg)

Range: Middle East, northern India, and Saharan Africa

Habitat: Deserts

Diet: Desert vegetation, short grasses, and plants

PRZEWALSKI'S HORSE

(EQUUS FERUS PRZEWALSKII)

Przewalski's horses are the only true wild horses left on Earth and are critically endangered. They are small and stocky. The hair on their manes and tails sheds and regrows annually. Each year, their coats grow thick and warm to withstand the harsh Mongolian winters.

HOW TO SPOT

Size: 4 to 4.6 feet (1.2 to 1.4 m) at the shoulder; 440 to 660 pounds (200 to 300 kg)

Range: Mongolia, Kazakhstan, and northern China

Habitat: Grasslands

Diet: Wild grasses and leaves

FUN FACT

In Mongolia, the Przewalski's horse is called *takhi*, which means "spirit."

GREVY'S ZEBRA *(EQUUS GREVYI)*

There are three species of zebra, which are part of the horse family. The Grevy's zebra is the largest and stockiest of all the zebras. It also has the largest ears. Like all zebras, the Grevy's zebra has black and white stripes. These are like fingerprints—each stripe pattern is unique to an individual. The stripes are a form of protection. When zebras stand together, it is difficult for predators to tell where one zebra ends and the other one starts. This makes it hard for predators to zero in on one animal.

HOW TO SPOT

Size: 4 to 5 feet (1.2 to 1.5 m) at the shoulder; 770 to 950 pounds (350 to 430 kg)

Range: Northern Kenya and southern Ethiopia

Habitat: Grasslands and savannas

Diet: Grasses, leaves, fruit, and bark

MALAYAN TAPIR *(TAPIRUS INDICUS)*

There are four species of tapirs, and Malayan tapirs are the largest. Because of their snouts, tapirs look like pigs, but they are more closely related to horses and rhinos. All tapirs have a short trunk used for gripping. They use this to pluck fruit or to strip leaves off branches. The Malayan tapir is easily distinguished from other tapirs by its distinct, black front half and rear legs and its white middle.

HOW TO SPOT

Size: 3 to 3.6 feet (0.9 to 1.1 m) at the shoulder; 550 to 1,190 pounds (250 to 540 kg); females are slightly larger than males

Range: Southeast Asia

Habitat: Tropical rain forests and swamps

Diet: Leaves, buds, bark, fruit, grasses, and aquatic plants

SUMATRAN RHINOCEROS

(DICERORHINUS SUMATRENSIS)

The Sumatran rhinoceros is the smallest of the rhino species. They are short and stocky, with two horns on their snouts. The front horn is much larger than the other. Sumatran rhinos have dark-gray to brown, leathery, thick skin with deep folds. Their skin also has short, stiff hairs that help keep mud trapped on their bodies. This protects the rhinos from insects and keeps them cool. These rhinos tend to be solitary, coming together only to mate.

HOW TO SPOT

Size: 3.7 to 4.8 feet (1.1 to 1.5 m) at the shoulder; 1,760 to 4,400 pounds (800 to 2,000 kg)

Range: Southeast Asia

Habitat: Rain forests and scrub forests

Diet: Leaves, shrubs, fruit, and twigs

FUN FACT

Ancestors of rhinos lived 55 million years ago. There have been almost 100 different rhino species since, with five species existing into the modern era. They are found only in Asia and Africa.

AFRICAN ELEPHANT

(LOXODONTA AFRICANA)

African elephants are the largest land animals on Earth. Both males and females have two curved ivory tusks that can be as long as 11.5 feet (3.5 m). The tusks are used to dig for food or to strip bark from trees. Males also use their tusks to fight one another. Female African elephants live in a herd that can range in size from only a few to dozens. Other members of the herd include young. Once they are old enough to go out on their own, males, called bulls, live mostly alone or with a few other males.

HOW TO SPOT

Size: 8.2 to 13 feet (2.5 to 4 m) at the shoulder; 5,000 to 14,000 pounds (2,300 to 6,400 kg)

Range: Central and southern Africa

Habitat: Savannas, woodlands, and scrub forests

Diet: Leaves, roots, grasses, fruit, and bark

FUN FACT

Elephants' trunks are used for many purposes: drinking, grabbing, breathing, communicating, and sniffing. They can also use their trunks like snorkels when they swim across deep water. Their trunks have more than 100,000 different muscles.

ASIAN ELEPHANT *(ELEPHAS MAXIMUS)*

Asian elephants are considered forest elephants, although they are seen at the edges of grasslands as well. The grasslands provide a better food source for the elephants, while the forests provide a way to escape from danger and to protect themselves from the hot sun. Like African elephants, Asian elephants live in small herds of females and their young. These adult females will work together to raise young and protect the group. Only some male Asian elephants grow long tusks.

WHAT'S THE DIFFERENCE?

Asian and African elephants are two distinct species. African elephants are larger. Their ears are also larger and somewhat resemble the shape of Africa. The ears of Asian elephants are smaller and rounder. The head shape of African elephants is rounder and dome-like. Asian elephants have heads with two domes and a dent in between. The skin of African elephants is much more wrinkly than that of Asian elephants.

HOW TO SPOT

Size: 6.6 to 9.8 feet (2 to 3 m) at the shoulder; 4,500 to 11,000 pounds (2,000 to 5,000 kg)

Range: India and Southeast Asia

Habitat: Mostly forested tropical regions and some grasslands

Diet: Grasses, leaves, roots, fruit, twigs, and bark

RED KANGAROO *(MACROPUS RUFUS)*

Red kangaroos are the largest of all marsupials. The males have a reddish-brown coat, while females' coats are bluish gray. Females have a pouch on their abdomens for carrying and nursing young. The young are called joeys. All kangaroos have powerful hind legs, as well as a powerful tail. The tail is used for balance. Males will lean back on their tails to fight other males with their hind legs. With their strong legs, red kangaroos can hop a distance of 25 feet (7.6 m) in one jump and reach a height of 6 feet (1.8 m). They can also reach speeds of more than 35 miles per hour (56 kmh).

HOW TO SPOT

Size: Head and body length is 3.25 to 5.25 feet (1 to 1.6 m); tail length is 3 to 3.6 feet (0.9 to 1.1 m); up to 200 pounds (90 kg)

Range: Central and inland Australia

Habitat: Deserts, open grasslands, and scrublands

Diet: Grasses and flowering plants

Male

WHAT IS A MARSUPIAL?

Marsupials are pouched mammals. That means the females have a pouch on their bodies to carry their young. Some young marsupials stay in the pouch for months. Once they are bigger and more developed, they leave the pouch and cling to their mothers' fur.

A female with a joey in her pouch

BENNETT'S WALLABY

(MACROPUS RUFOGRISEUS)

All wallabies are smaller than kangaroos, but like kangaroos, wallabies have powerful hind legs and a tail. Bennett's wallabies have light-gray fur, a white stomach, and darker paws and muzzles, as well as large ears compared to other similar species. They also have a hint of reddish-brown fur on their backs at the shoulders and neck. Because of this, they are sometimes called red-necked wallabies. Their main mode of movement is hopping, and they are able to reach speeds of 9 miles per hour (14.5 kmh).

FUN FACT

The kangaroo family includes kangaroos, wallabies, wallaroos, tree kangaroos, and more. There are more than 50 species in this family, whose scientific name, *Macropodidae*, means "big feet."

HOW TO SPOT

Size: 3 to 3.5 feet (0.9 to 1.1 m) tall; tail length is 2.3 to 2.4 feet (0.7 to 0.73 m); 30 to 40 pounds (14 to 18 kg)

Range: Tasmania and the east coast of Australia

Habitat: Eucalyptus forests and grasslands

Diet: Grasses, herbs, and roots

COMMON WOMBAT

(VOMBATUS URSINUS)

Common wombats are short and pudgy with a wide, round head and stubby tail. They are grayish brown. They have strong claws for digging burrows, where they live. They use their rear ends to bulldoze the loose dirt out of a tunnel. Sometimes there are many connected tunnels and chambers within the burrow. Wombats live in the borrows alone. Common wombats may spend up to 16 hours a day resting in their burrows. They come out at dusk to graze.

HOW TO SPOT

Size: 2.3 to 3.9 feet (0.7 to 1.2 m) tall; 32 to 80 pounds (14.5 to 36 kg)

Range: Eastern Australia and several nearby islands

Habitat: Temperate forests and grasslands

Diet: Grasses, bark, roots, and shrubs

KOALA *(PHASCOLARCTOS CINEREUS)*

Koalas have grayish fur and round, fuzzy ears. They spend almost their entire lives in eucalyptus trees. When not eating, they rest in the crook of two branches. They curl their feet and hands around a branch and use their sharp claws to dig into the bark. Koalas nap more than 18 hours a day. That's because koalas don't get a lot of energy from their food.

HOW TO SPOT

Size: 2 to 3 feet (0.6 to 0.9 m) tall; 9 to 29 pounds (4 to 13 kg)

Range: Eastern and southeastern Australia

Habitat: Eucalyptus forests

Diet: Eucalyptus leaves

FUN FACT

Koalas have a special claw for grooming. They use this to clean themselves instead of licking, which is what many other mammals do.

VIRGINIA OPOSSUM

(DIDELPHIS VIRGINIANA)

These opossums are the only marsupials found in North America. They waddle when they walk but are excellent climbers. They use their tails for gripping and balance. Virginia opossums are highly adapted to a variety of habitats, including areas inhabited by humans. They have grayish fur with a lighter face and a long, pointed snout. The snout sniffs out a wide variety of both plant and animal food sources, as well as garbage left by humans.

HOW TO SPOT

Size: 14 to 20 inches (36 to 51 cm); tail length is 9 to 15 inches (23 to 38 cm); 2 to 15 pounds (1 to 7 kg)

Range: Central and eastern North America

Habitat: Fields, wetlands, and forests; urban and suburban areas

Diet: A variety of fruit and plants, insects, small animals, and carrion

TASMANIAN DEVIL

(SARCOPHILUS HARRISII)

Tasmanian devils are known for their bad tempers. When they are threatened, protecting their food, or fighting for a mate, devils will fly into a rage that includes baring their teeth, screaming, snarling, and lunging. Tasmanian devils are stocky with blackish-brown fur. They also have a single white, horizontal stripe across their chests and shoulders. Tasmanian devils use their excellent senses of smell to find food. They have sharp teeth and powerful jaws to eat their meals—bones and all.

FUN FACT

Tasmanian devils have the nickname "vacuum cleaners of the forest." They got this name because they eat animals that have died—cleaning up the forest as they eat.

HOW TO SPOT

Size: 23 to 26 inches (58 to 66 cm); tail length is 9 to 10 inches (23 to 25 cm); 11 to 30 pounds (5 to 14 kg)

Range: Tasmania, an island near Australia

Habitat: Forests, scrublands, and fields

Diet: Carrion; sometimes small birds or mammals

PLATYPUS *(ORNITHORHYNCHUS ANATINUS)*

Platypuses are one of Earth's most unusual animals. They look part beaver because of their tails. Their bills and webbed feet make them look part duck. And they seem to be part otter because of their body shapes and dense, waterproof fur. Plus, unlike almost all other mammals, they lay eggs. For defense, males also have a sharp, venomous spur on their ankles. To find food in murky water, platypuses rely on their sensitive bills to sense electrical signals from potential prey.

HOW TO SPOT

Size: 15 inches (38 cm); tail length is 5 inches (13 cm); 3 pounds (1.4 kg)

Range: Eastern and southeastern Australia; Tasmania

Habitat: Wetland areas including rivers, streams, lagoons, and ponds

Diet: Insect larvae, worms, and shellfish

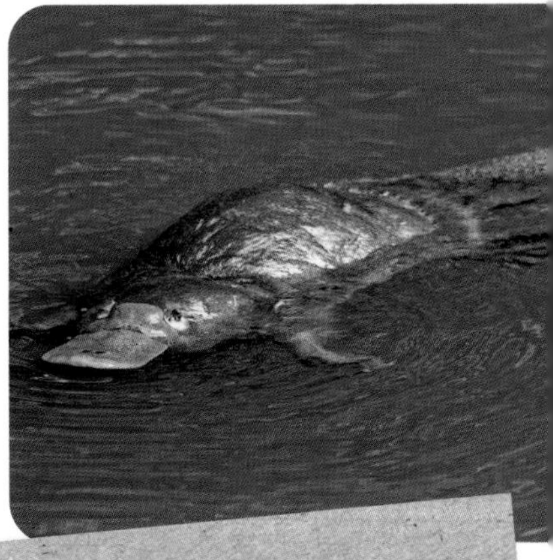

FUN FACT

When a platypus specimen was first sent to Europe from Australia in the late 1700s, people thought it was a joke. That's because the mammal looks so strange.

SHORT-BEAKED ECHIDNA

(TACHYGLOSSUS ACULEATUS)

The short-beaked echidna is covered with spines. The spines are mostly yellow and are tipped with black. These creatures use their claws and tube-like snouts to dig into termite and ant nests. Then they use their sticky tongues to snatch up prey. Female short-beaked echidnas carry their young inside a pouch. When the young begin to grow spines, the mother will take them out and put them in a burrow.

HOW TO SPOT

Size: 17.3 to 22 inches (44 to 55 cm) long; 4.4 to 15 pounds (2 to 7 kg)

Range: Australia

Habitat: Woodlands, forests, grasslands, and dry environments

Diet: Termites and ants

MONOTREMES

Monotremes do not give birth to live young. They lay eggs. They also have extremely sensitive bills or snouts that they use to detect prey, and they don't have teeth. There are only five species of monotremes: the platypus and four species of echidna.

GLOSSARY

adaptation
A change that allows something to become better modified to its surroundings.

aquatic
Living or growing in water.

baleen plates
The material in a baleen whale's mouth that acts like a strainer to trap food.

camouflage
The coloring or pattern of an animal that helps it blend into its surroundings.

carnivore
An animal that primarily eats meat.

carrion
The dead and decomposing meat of an animal.

domesticated
Animals that are no longer wild; those that are kept as pets or for food or work.

echolocation
The use of sounds to find objects such as prey.

herbivore
An animal that only eats plants.

insectivore
An animal that eats mostly insects.

invertebrate
An animal that does not have a backbone.

omnivore
An animal that eats both meat and plants.

savanna
A flat, grassy plain with few trees.

sub-Saharan
The part of Africa south of the Sahara Desert.

subspecies
A group of animals that are a division of the main species.

vertebrate
An animal with a backbone.

vocalize
To make a sound to communicate.

TO LEARN MORE

FURTHER READINGS

Alonso, Juan Carlos. *Marine Animals of the World*. Walter Foster, 2018.

Daniels, Patricia. *Mammals*. National Geographic, 2019.

Edwards, Sue Bradford. *The Evolution of Mammals*. Abdo, 2019.

ONLINE RESOURCES

To learn more about mammals, please visit **abdobooklinks.com** or scan this QR code. These links are routinely monitored and updated to provide the most current information available.

PHOTO CREDITS

Cover Photos: Daria Rybakova, bear; Shutterstock Images, dolphin, gray wolf, hedgehog, koala, lemur, elephant, back (Virginia opossum); Eric Isselle/Shutterstock Images, cheetah, raccoon; Jeff McGraw/Shutterstock Images, caribou; Jim Cumming/Shutterstock Images, snowshoe hare; Ivan Kuzmin/Shutterstock Images, back (bat); iStockphoto, back (skunk); Andrew Haysom/iStockphoto, back (echidna)

Interior Photos: iStockphoto, 1 (capybara), 1 (panda), 1 (tiger), 6 (bottom), 9 (top), 9 (bottom), 10, 12 (bottom), 13, 16 (top), 16 (bottom), 18, 19 (top), 21 (top), 21 (bottom), 22 (bottom), 23, 24, 25 (top), 25 (bottom), 28, 29 (top), 29 (bottom), 30, 31 (top), 32 (bottom), 34 (top), 38 (top), 38 (bottom), 39, 41, 43 (top), 43 (bottom), 45 (bottom), 47, 49 (top), 50 (bottom), 52 (top), 53, 54, 55 (bottom), 56 (top), 56 (bottom), 60, 61 (bottom), 64 (top), 64 (bottom), 66, 72, 78, 79 (top), 79 (bottom), 82 (bottom), 84, 85 (top), 88 (top), 89, 90, 91 (bottom), 92 (right), 93, 94, 95 (top), 95 (bottom), 96 (top), 96 (bottom), 98, 100 (left), 102 (top), 102 (bottom), 105 (top), 112 (mandrill); Andrew Haysom/iStockphoto, 1 (echidna), 107; Media Production/iStockphoto, 1 (raccoon); Eric Isselle/Shutterstock Images, 4 (cheetah), 5 (echidna), 62; Shutterstock Images, 4 (gray wolf), 5 (dolphin), 5 (koala), 5 (elephant), 11 (top), 34 (bottom), 44, 75, 82 (top), 104 (top), 112 (Virginia opossum), 112 (African dog); Daria Rybakova, 5 (bear); Gray Photo Online/Shutterstock Images, 5 (chipmunk); Jim Kruger/iStockphoto, 6 (top), 32 (top); Sophia Granchinho/Shutterstock Images, 7 (top), 33 (top); Natures Moments UK/Shutterstock Images, 7 (bottom), 33 (bottom); Doc Whiteantheo / Pantheon, 8; Mark Caunt/Shutterstock Images, 11 (bottom); Nicolas Me/iStockphoto, 12 (top); Benny Marty/iStockphoto, 14; Jeremy Richards/iStockphoto, 15 (top), 15 (bottom); Sepp Friedhuber/iStockphoto, 17; Laura Din/iStockphoto, 19 (bottom); Keith Binns/iStockphoto, 20, 112 (black bear); Nuno Tendais/iStockphoto, 22 (top); Mike Lane/iStockphoto, 26 (top), 77 (bottom), 86 (top); John Pitcher/iStockphoto, 26 (bottom); Anta Elder Design/iStockphoto, 27; June Jacobsen/iStockphoto, 31 (bottom); John Carnemolla/Shutterstock Images, 35; Gilbert S. Grant/Science Source, 36 (top); Dominic Gentilcore/Shutterstock Images, 36 (bottom); Robert Winkler/iStockphoto, 37; Steven Schremp/iStockphoto, 40 (top); Blue Barron Photo/iStockphoto, 40 (bottom); Mint

Images/SuperStock, 42 (top); Neil Bromhall/Shutterstock Images, 42 (bottom); M. Leonard Photography/iStockphoto, 45 (top); Lynn Bystrom/iStockphoto, 46 (top); Evgenii Mitroshin/ Shutterstock Images, 46 (bottom); Elliotte Rusty Harold/ Shutterstock Images, 48; John Randall Alves/iStockphoto, 49 (bottom); Holly Kuchera/ Shutterstock Images, 50 (top); Paul Reeves Photography/ Shutterstock Images, 51; Denisa Pro/Shutterstock Images, 52 (bottom); Media Production/ iStockphoto, 55 (top); Anan Kaewkhammul/Shutterstock Images, 57, 112 (meerkat); Justin Davis/iStockphoto, 58; Dan Guravich/Science Source, 59 (top); Helissa Grundemann/ Shutterstock Images, 59 (bottom); Pascale Gueret/ iStockphoto, 61 (top); FLPA/ SuperStock, 63 (top); Rod Planck/ Science Source, 63 (bottom); Phil Degginger/Science Source, 65; Foto Mous/Shutterstock Images, 67 (top); Biosphoto/SuperStock, 67 (bottom); Steve Downeranth/ Pantheon/SuperStock, 68; Michael Durham/Minden Pictures/Newscom, 69; Ivan Kuzmin/Shutterstock Images, 70 (left), 70 (right), 104 (bottom); Animals Animals/ SuperStock, 71; Natalia Kuzmina/ Shutterstock Images, 73 (top), 73 (bottom); David Thyberg/ Shutterstock Images, 74 (top); Tony Camacho/Science Source, 74 (bottom); Dave Hamilton/ iStockphoto, 76; THP Stock/ iStockphoto, 77 (top); USO/ iStockphoto, 80 (top); Jeff W. Jarrett/Shutterstock Images, 80 (bottom); Jeff McCurry/ iStockphoto, 81; Dennis van de Water/Shutterstock Images, 83; Chuck Schug Photography/ iStockphoto, 85 (bottom); Martin Mecnarowski/Shutterstock Images, 86 (bottom); Harry Collins/iStockphoto, 87; Daniel Gaura/iStockphoto, 88 (bottom); VDCM image/iStockphoto, 91 (top); Tracie Louise/iStockphoto, 92 (left); Terry Whittaker/Science Source, 97; Rudy Umans/ Shutterstock Images, 99; John Carnemolla/iStockphoto, 100 (right), 105 (bottom), 106 (bottom); Lakeview Images/iStockphoto, 101; Alizada Studios/iStockphoto, 103; Martin Pelanek/Shutterstock Images, 106 (top)

ABDOBOOKS.COM
Published by Abdo Publishing, a division of ABDO, PO Box 398166, Minneapolis, Minnesota 55439.

Printed in the United States of America, North Mankato, Minnesota.
082020
012021

Editor: Alyssa Krekelberg
Series Designer: Colleen McLaren

Library of Congress Control Number: 2019954383
Publisher's Cataloging-in-Publication Data
Names: Perdew, Laura, author.
Title: Mammals / by Laura Perdew
Description: Minneapolis, Minnesota : Abdo Publishing, 2021 | Series: Field guides for kids | Includes online resources and index.
Identifiers: ISBN 9781532193064 (lib. bdg.) | ISBN 9781098210960 (ebook)
Subjects: LCSH: Mammals--Juvenile literature. | Mammals--Behavior--Juvenile literature. | Animals--Field guides--Juvenile literature. | Reference materials--Juvenile literature.
Classification: DDC 599--dc23